DEVELOPMENTS IN
FOOD PRESERVATION—3

CONTENTS OF VOLUMES 1 AND 2

Volume 1

1. Technology Appropriate to Food Preservation in Developing Countries. MARC R. BACHMANN

2. Cooling of Horticultural Produce with Heat and Mass Transfer by Diffusion. G. VAN BEEK and H. F. TH. MEFFERT

3. The Preparation of Fruit Juice Semi-concentrates by Reverse Osmosis. M. DEMECZKY, M. KHELL-WICKLEIN and E. GODEK-KERÉK

4. The Effect of Microwave Processing on the Chemical, Physical and Organoleptic Properties of Some Foods. MARGARET A. HILL

5. Freeze Drying: The Process, Equipment and Products. J. LORENTZEN

6. Extrusion Processing—A Study in Basic Phenomena and Application of Systems Analysis. JUHANI OLKKU

7. The Effect of Temperature on the Deterioration of Stored Agricultural Produce. JOSÉ SEGURAJAUREGUI ALVAREZ and STUART THORNE

8. Thermal Sterilisation of Foods. L. W. WILLENBORG

Volume 2

1. Controlled Atmosphere Storage of Fruits and Vegetables. D. H. DEWEY

2. Food Irradiation. J. F. DIEHL

3. Heat and Mass Transport in Solid Foods. B. HALLSTRÖM and C. SKJÖLDEBRAND

4. Recent Developments in Spray Drying. KEITH MASTERS

5. Computers in Food Processing. A. ROSENTHAL

6. Ethylene in the Storage of Fresh Produce. ZULEIKHA TALIB

DEVELOPMENTS IN FOOD PRESERVATION—3

Edited by

STUART THORNE

Department of Food and Nutritional Sciences,
King's College, University of London, UK

ELSEVIER APPLIED SCIENCE PUBLISHERS
LONDON and NEW YORK

ELSEVIER APPLIED SCIENCE PUBLISHERS LTD
Crown House, Linton Road, Barking, Essex IG11 8JU, England

Sole Distributor in the USA and Canada
ELSEVIER SCIENCE PUBLISHING CO., INC.
52 Vanderbilt Avenue, New York, NY 10017, USA

British Library Cataloguing in Publication Data

Developments in food preservation.
3
1. Food—Preservation
I. Title
641.4   TP371.2

ISBN 0-85334-384-5

WITH 33 TABLES AND 54 ILLUSTRATIONS

© ELSEVIER APPLIED SCIENCE PUBLISHERS LTD 1985

Printed in Northern Ireland at The Universities Press (Belfast) Ltd.

PREFACE

A major trend in food preservation is towards the more efficient use of energy. The main initiative for this has been the rapid rise in energy costs over the past decade. A dozen years ago, the energy costs of food processing and preservation operations were often considered as an inevitable and fixed cost of the process. Today, however, they make a significant contribution to the total cost of food processing and reduction of energy use is vital to the financial well-being of the food industry. The optimisation of the thermal behaviour of food processing plant involves mathematical modelling of the system and, later, experimental verification of the model. The equations involved in such models, particularly where unsteady state heat and mass transfer are involved, are complex. We are fortunate that the need for such models has coincided with the general availability of powerful computers, without which numerical solutions to performance equations would be difficult if not impossible to obtain. The result of all this activity aimed at improving the efficiency of food processing operations is that there has been a very rapid increase in our fundamental understanding of food processing operations in general, and not just thermal aspects. Improvement in the understanding of processes inevitably leads to improvement in the application of the processes and better product quality.

Many food processes rely for their effectiveness on the attainment of prescribed temperature–time relationships and improvement in thermal efficiency is often synonymous with the establishment of the best temperature–time relationship from the point of view of product quality. Food processing is rapidly changing for the better; from an art

v

to a scientifically based technology. This must result in an improved food industry if we can induce those responsible for formulation of products to eliminate some of the excessive and unnecessary use of additives and 'artificial' ingredients that characterise a few products of the food industry and bring the rest of the industry into disrepute.

This third volume of *Developments in Food Preservation* concentrates on some of the major improvements in the fundamental understanding of food preservation processes in recent years and, in particular, on improvements in the thermal efficiency of processes. Two chapters (5 and 6) are concerned directly with the thermal performance of heat exchangers. Chapter 4 is concerned with the application of high frequency energy for heating as an alternative method for transferring heat to foods, often more efficient than conventional heating by conduction, convection and radiation. Chapter 1, although not directly involved with thermal efficiency, discusses the current state of knowledge of the thermal destruction of bacterial spores, a proper knowledge of which reduces the over-cooking characteristic of many thermally sterilised foods and leads, indirectly, to thermally more efficient operation.

Of the remaining, two chapters (2 and 3) deal with vital aspects of food hygiene and microbiology, both essential to the effective and efficient prosecution of food preservation. Cleaning of food processing plant is not only essential to the safe operation of processes, but also contributes to the efficient operation of the plant. Packaged meats (Chapter 2) are products of increasing commercial importance with an extremely complex microbial flora, which must be controlled to provide a safe and acceptable product. The last chapter (Chapter 7) is a survey of the state of the art of fluidisation in food processing, a process that has found some application for several decades. The high heat and mass transfer coefficients that can be achieved in fluidised beds offer great potential for its application to many other processes in the food industry.

STUART THORNE

CONTENTS

Preface v

List of Contributors ix

1. Determination of Thermal Processes to Ensure Commercial
   Sterility of Foods in Cans 1
   ANDREW C. CLELAND and GORDON L. ROBERTSON

2. Refrigerated Storage of Packaged Meat 45
   H.-J. S. NIELSEN

3. Cleaning of Food Processing Plant 95
   ALAN T. JACKSON

4. Application of Dielectric Techniques in Food Production and
   Preservation 127
   MIHÁLY DEMECZKY

5. Design and Optimisation of Falling-Film Evaporators . . . 183
   MAURO MORESI

6. Heat Transfer and Sterilisation in Continuous Flow Heat
   Exchangers 245
   CHRISTIAN TRÄGÅRDH and BERT-OVE PAULSSON

7. Potential Applications of Fluidisation to Food Preservation . 273
   GILBERT M. RIOS, HENRI GIBERT and JEAN L. BAXERRES

Index 305

LIST OF CONTRIBUTORS

JEAN L. BAXERRES
GEPICA, *Institut du Génie Chimique, Toulouse, France*

ANDREW C. CLELAND
Department of Biotechnology, Massey University, Palmerston North, New Zealand

MIHÁLY DEMECZKY
Central Food Research Institute, Herman Ottó út 15, H-1525 Budapest, Hungary

HENRI GIBERT
GEPICA, *Institut du Génie Chimique, Toulouse, France*

ALAN T. JACKSON
Department of Chemical Engineering, University of Leeds, Leeds LS2 9JT, UK

MAURO MORESI
Dipartimento di Ingegneria Chimica, Materiali, Materie Prime e Metallurgia, Università degli Studi di Roma—La Sapienza, Via Eudossiana 18, I-00184 Rome, Italy

H.-J. S. NIELSEN
Laboratoriet for Levnedsmiddelindustri, Danmarks Tekniske Hojskole, Bygning 221, 2800 Lyngby, Denmark

BERT-OVE PAULSSON
Division of Food Engineering, Lund University, PO Pox 124, S-221 00 Lund, Sweden

GILBERT M. RIOS

Laboratoire de Génie Chimique, Université des Sciences et Techniques du Languedoc, Place Eugène Bataillon, 34060 Montpellier Cedex, France

GORDON L. ROBERTSON

Department of Food Technology, Massey University, Palmerston North, New Zealand

CHRISTIAN TRÄGÅRDH

Division of Food Engineering, Lund University, PO Box 124, S-221 00 Lund, Sweden

Chapter 1

DETERMINATION OF THERMAL PROCESSES TO ENSURE COMMERCIAL STERILITY OF FOOD IN CANS

Andrew C. Cleland and Gordon L. Robertson

Departments of Biotechnology and Food Technology, Massey University, Palmerston North, New Zealand

SUMMARY

Application of heat to food in sealed containers is the most common method of obtaining a commercially sterile product. Methods available for the calculation of the necessary heating conditions are reviewed. Weaknesses exist in the assumption that spore inactivation is logarithmic, and in a number of the mathematical models used to estimate temperature/time histories (and hence F values) during processing. Uncertainties in data and shortcomings in methods to account for these in calculations decrease markedly the precision with which process lethality can be determined. As well as attempting to make cannery technologists and regulatory authorities more aware of the limitations in the techniques currently used commercially, the review is intended to provide some insight into areas in which future research might be most useful.

NOMENCLATURE

a_w Water activity
A Frequency factor in Arrhenius relationship (s^{-1} or min^{-1})
b Coefficient in eqn (10) ($°C^{-1}$)
C Concentration of spores (m^{-3} or ml^{-1})
C_f Concentration of survivors following processing (m^{-3} or ml^{-1})

1

C_{fc} Concentration of survivors at the critical point (m^{-3} or ml^{-1})

C_0 Initial concentration of spores (m^{-3} or ml^{-1})

D Time for a 10-fold reduction in spore numbers (s or min)

D_r D evaluated at temperature θ_r (s or min)

$D_{121\cdot1}$ D evaluated at 121·1°C (s or min)

E_a Activation energy (J mol^{-1})

E_1 Exponential integral

f Reciprocal of slope of logarithmic temperature difference vs time plot (s or min)

f_c f value for cooling (s or min)

f_h f value for heating (s or min)

F Lethality of a process defined by equivalent time at some temperature for a specified z value (s or min)

F_c F value for the cooling phase (s or min)

F_h F value for the heating phase (s or min)

F_o F based on a reference temperature of 121·1°C and $z = 10$°C (s or min)

F_r F determined at temperature θ_r (s or min)

F_{ra} F_r value determined using the Arrhenius relationship (s or min)

F_{rl} F_r value determined using the linear temperature model (s or min)

F_t Total F for a process (s or min)

H Can height (m)

i Parameter used in numerical integration of L values

j_c Dimensionless intercept parameter for heat transfer in cooling

j_h Dimensionless intercept parameter for heat transfer in heating

k Rate constant for spore inactivation (s^{-1} or min^{-1})

k_r Rate constant at temperature θ_r (s^{-1} or min^{-1})

L Dimensionless lethal rate parameter

N Number of spores in a container

N_f Number of spores surviving a process in a container

N_0 Initial number of spores in a container

r Displacement from can centre in radial direction (m)

Δr Space step used in numerical integrations (m)

r_c Can radius (m)

R Gas constant (J $mol^{-1}\,K^{-1}$)

t Time (s or min)

t_h Process heating time (s or min)

t_t Total process time (s or min)

t_{ta} Total process time determined using Arrhenius relationship (s or min)

t_{tl} Total process time determined using linear temperature model (s or min)

Δt Time step in numerical integrations (s or min)

T Absolute temperature (K)

T_r Reference absolute temperature (K)

V Volume (m^3 or ml)

V_c Can volume (m^3 or ml)

x Displacement from can centre in axial direction (m)

Δx Space step used in numerical integrations (m)

z Temperature interval for a 10-fold change in k (°C)

α Thermal diffusivity of food in can (m^2 s^{-1})

θ Temperature (°C)

θ_c Water temperature in cooling (°C)

θ_{fc} Temperature at critical point in can (°C)

θ_g Temperature at critical point at end of heating phase (°C)

θ_h Steam temperature in heating (°C)

θ_{in} Initial temperature of can contents (°C)

θ_r Reference temperature (°C)

1. PRINCIPLES OF THERMAL PROCESSING

1.1. Introduction

The thermal processing of foods is a common operation in the food industry, ranging from the relatively mild treatments used in pasteurisation and blanching through cooking to the relatively severe treatments necessary to achieve sterilisation. Although an increasing quantity of food is sterilised prior to being packaged aseptically, the majority of food is still packaged in either metal cans or flexible pouches and then sterilised. It is this last operation, commonly referred to as canning, which is the focus of this review, although many of the principles are equally applicable to the thermal processes of pasteurisation, blanching or cooking.

The criterion of success in the canning process is the inability of the micro-organisms and their spores to grow under conditions normally encountered in storage. This means that there could be some dormant non-pathogenic micro-organisms in the food (any pathogenic organisms are dead) but environmental conditions are such that the organisms do not reproduce. Foods which have been thermally processed with this as the criterion are referred to as 'commercially sterile'. Although it has been suggested that the term should be dismissed for semantic reasons[1] it has been adopted by Codex Alimentarius,[2] which has defined the commercial sterility of thermally processed foods as the condition achieved by the application of heat, sufficient, alone or in combination with other appropriate treatments, to render the food free from micro-organisms capable of growing in the food at normal non-refrigerated conditions at which the food is likely to be held during distribution and storage.

The thermal conditions needed to produce commercial sterility depend on many factors including

(a)   the nature of the food (e.g. pH and a_w);
(b)   storage conditions of the food following the thermal process;
(c)   heat resistance of the micro-organisms or spores;
(d)   heat transfer characteristics of the food, its container and the heating medium and
(e)   the initial load of micro-organisms.

An optimal thermal process is the minimum treatment that achieves commercial sterility because both heating cost and product quality loss increase as a process time is lengthened.

The cylindrical can is the major type of sealed container used, and the most common method of processing is the application of steam under pressure followed by cooling in water. Research results published in the literature have therefore concentrated on this container and process. This review will consider literature relevant to these, and define aspects worthy of further investigation.

1.2. Definition of Sterility

The concept normally used to define sterility is that only one in a large number of cans would be contaminated after processing, and that with a single viable spore.[3] The average number of spores in a can (N) is therefore less than unity. N has normally been considered as the probability that the can has a spore surviving in it. Mathematically

$$N = \int_0^{V_c} C \, dV \qquad (1)$$

At the commencement of a thermal process the number of spores in a can is denoted N_0, and the number/probability at the end of the process N_f. The value that N_f should never exceed for a product to be commercially sterile is ill-defined. The $12D$ concept is often referred to. The parameter D designates the time for a 10-fold reduction in spore numbers at a specified temperature. Although Stumbo[3,4] suggests that a $12D$ process is one for which

$$N_f \leq 10^{-12} \tag{2}$$

the alternative interpretation

$$N_f/N_0 \leq 10^{-12} \tag{3}$$

(e.g. Pflug and Odlaug[5]) is more common. Equation (2) suggests that 1 in 10^{12} cans will be contaminated, whereas eqn (3) implies that a greater fraction will be contaminated if N_0 exceeds 1. An N_0 value of 10^6 would mean that a spore would survive in one can in a million. Pflug and Odlaug[5] suggest that appropriate N_f values might be 10^{-9} for pathogenic organisms, and 10^{-6} for non-pathogenic spoilage organisms. There appears to be no standard N_f value in use.

The total lethality or sterilising value is usually designated by the symbol F. A subscript on F indicates the reference temperature, and a superscript the z value (z is a parameter which characterises the relative resistance of spores of a specific organism to temperature). F values based on a reference temperature of $121.1°C$ ($250°F$) and a z value of $10°C$ ($18°F$) are designated F_o.

From the beginning F has embodied two different concepts. One concept views F as the goal or requirement of the process; it is determined by the relationship

$$F = D(\log N_0 - \log N_f) \tag{4}$$

The choice of D is determined by the target organism or the objective of the process. Tables of D values for organisms of interest to canners are widely available (e.g. Stumbo[3]).

The second concept of F views it as a measure of the accomplishment of an actual process (methods for calculation of this F are discussed in Section 2). Although the two values for F are arrived at by different calculations, they should be equal.

1.3. Commercial Practice

1.3.1. Safety

For a low-acid canned food (pH $\geqslant 4\cdot5$) the minimum process must assure safety by destroying any contaminating *Clostridium botulinum*. This is normally considered to be accomplished by a $12D$ process. Using the definition of eqn (3), if N_0 is 10^3, a $12D$ process will result in an N_f value of 10^{-9}, or the expectation of one container in 10^9 containing a survivor.

Stumbo[3] states that for *C. botulinum* the most resistant spores are characterised by $D_{121\cdot1} = 0\cdot21$ min and $z = 10°C$. Therefore eqn (4) defines a $12D$ process for *C. botulinum* as one with a sterilising value (F_o) of $(0\cdot21) \times (12)$ or $2\cdot52$ min, which is conventionally rounded off to 3 min. Using the less common definition of eqn (2) the F_o value to achieve $N_f = 10^{-12}$ from $N_0 = 10^3$ is $3\cdot15$ min. The F_o value for a $12D$ process, however determined, is usually taken as the processing goal for safety.

In some cases the $12D$ process defined by eqn (3) may prove inadequate. For example, a 450 ml can with 10^4 spores ml^{-1} would have a probability of spoilage of $4\cdot50 \times 10^{-6}$ if an $F_o = 2\cdot52$ min process was used. A better approach would be to define an acceptable N_f value (e.g. $\leqslant 10^{-9}$) and calculate the minimum safe process from eqn (4).

Data for *C. botulinum* outbreaks[5] show that the failure of the operator to deliver the specified process has been the most common cause. Incorrect process calculation was less important.

1.3.2. Spoilage

The value of D for a typical mesophilic spoilage organism is almost five times greater than that of *C. botulinum*; for example *Clostridium sporogenes* has a $D_{121\cdot1}$ value of $1\cdot0$ min.[3] The processor must decide what an acceptable level of spoilage is, and then calculate the F value for spoilage. For an initial contamination level of $N_0 = 10^3$ and the acceptance of one spoiled unit in every 1000 units, the F value would be:

$$F = (1\cdot0)(\log 10^3 - \log 10^{-3})$$

$$= 6\cdot0 \text{ min} \tag{5}$$

This F value is twice that needed for safety. A process goal that provides a reasonable assurance against spoilage is almost always more drastic than that necessary for safety. The extent of the safety margin

TABLE 1
VALUES OF F_o FOR SOME COMMERCIAL PROCESSES[6]

Product	Can sizes	F_o (min)
Asparagus	All	2–4
Green beans, brine packed	No. 2	3·5
Green beans, brine packed	No. 10	6
Chicken, boned	All	6–8
Corn, whole kernel, brine packed	No. 2	9
Corn, whole kernel, brine packed	No. 10	15
Cream style corn	No. 2	5–6
Cream style corn	No. 10	2·3
Dog food	No. 2	12
Dog food	No. 10	6
Mackerel in brine	301 × 411	2·9–3·6
Meat loaf	No. 2	6
Peas, brine packed	No. 2	7
Peas, brine packed	No. 10	11
Sausage, Vienna in brine	Various	5
Chili con carne	Various	6

Reprinted from Food Technology (1952) **6(2)**, 185. Copyright ©
by Institute of Food Technologists.

can be seen by reference to Table 1 which lists some F_o values used in commercial practice.[6] Guidelines such as Table 1 have been used successfully for many years although it is unlikely that all the data in the table have been derived by dual lethality calculations for pathogens and spoilage organisms.

Thermophilic spoilage is rarely the target for establishing a processing goal. The very high $D_{121·1}$ values (up to $4·0 \, min^3$) would result in a large sterilising value and the process needed to deliver the sterilising value would be excessively long. Thermophilic spoilage is not considered 'underprocessing', but rather the result of the improper cooling of cans or improper handling during distribution.[7] Further, surviving thermophiles would remain dormant in storage at room temperature.

1.3.3. Process Confirmation
The establishment of a thermal process should always involve two phases.[8] The first is the determination of the heating time at a specific retort temperature to achieve the required F. This involves heat penetration measurements and mathematical analysis of the data. The second phase is a follow-up test employing microbiological methods to

confirm the calculated process. This should be carried out wherever possible. There are two main methods of bacteriological process evaluation:

(a) The inoculated pack system, described in detail by the National Canners Association,[9] is basically a system for controlling the initial spore population and counting the number of survivors. By inoculating the can with a large number of spoilage organisms, the total population of typically 100 cans may become equivalent to that of 10 000 cans at normal contamination levels. Thus, the spoilage in 100 cans may approximate commercial operation. This method requires that the spore resistance be known for scaling the results of the tests.

(b) The count reduction system[8] involves inoculating test cans with $30-50 \times 10^6$ spores and running a series of tests of varying time and temperature. Survivor counts are determined on each processed can, and the D value of the organisms calculated from the survivor curves. The expected survivor concentration can then be calculated.

A further important aspect in designing the process is the way in which the cooling phase is handled. The steam may be shut off before the total target F is reached in the expectation that the cooling phase will increase the overall lethality to the required value. However, once the steam is off and cooling commenced, the lethal effect of the process is virtually fixed, although the contribution of the cooling phase to it is not yet known. Not unexpectedly there is reluctance to operate in this manner. The practice of reaching the total target F in the heating phase, and treating the cooling phase as a safety factor, is common in commercial practice.[10] This results in cans which are over-processed and have greater than necessary loss in heat-sensitive quality attributes. Further, processing costs are higher.

2. PRINCIPLES OF LETHALITY CALCULATIONS

In order to quantify operating conditions to achieve commercial sterility, mathematical models of three types are needed:

(a) a kinetic model for spore inactivation;
(b) a model for temperature effects on the kinetics and
(c) a model for heat transfer in the container.

The principles of the models available are discussed in this section. A critical analysis is made in Section 3.

2.1. Basic Kinetic Model

A first order kinetic model is almost invariably used for mathematical description of spore destruction as early work[11] suggested that it was probably an adequate description

$$dC/dt = -kC \tag{6}$$

$$C = C_0 \exp\left[-\int_0^t k \, dt\right] \tag{7}$$

These relationships are implicit in eqn (4) which introduced the idea of logarithmic destruction because

$$D = 2 \cdot 303/k \tag{8}$$

Use of eqn (7) in conjunction with eqn (1) allows N to be determined.

2.2. Effect of Temperature

The rate constant k, for the rate of thermal inactivation of spores is temperature dependent. Chemical kinetics commonly uses the Arrhenius relationship to describe the temperature dependence

$$k = A \exp(-E_a/RT) \tag{9}$$

This equation fits measured data for chemical reactions plausibly well. An approximation that can be applied over small temperature intervals is

$$k = k_r \exp[b(\theta - \theta_r)] \tag{10}$$

Historically the canning literature has used the latter although it is normally expressed as

$$k = k_r 10^{(\theta - \theta_r)/z} \tag{11}$$

A parameter called a lethal rate (L) is often used. It is defined as

$$L = k/k_r \tag{12}$$

It follows from eqn (8) that

$$L = D_r/D \tag{13}$$

The parameter z is the size of the temperature interval for a 10-fold change in k. Equation (11) implies that z remains constant with respect to temperature whereas the Arrhenius relationship yields

$$z = T/(E_a/2 \cdot 303RT + 1) \approx 2 \cdot 303RT^2/E_a \qquad (14)$$

2.3. Temperature/Time History in a Container

In all containers undergoing thermal processing the temperature varies with time, even if the external conditions such as steam temperature remain constant. In order to evaluate the integral in eqn (7) the temperature/time history at points of interest in the can must be known. There are two ways in which temperature profiles can be determined.

(a) Direct measurement, for example by placing thermocouples in a can.

(b) Predictions using heat transfer models such as pure convection or pure conduction. In the former the can contents are assumed to be perfectly mixed so that at any time there is no temperature difference between parts of the can. The practical applicability of such a model is limited. The pure conduction model assumes that throughout the process the can contents are completely immobilised. Heat conduction theory can then be used to calculate temperature/time profiles for any point in the can. To use either of the heat transfer models various heat transfer data are needed. These can only be found by experiment so the heat transfer models are no more than data extenders, being limited by any inaccuracy in the experiments conducted to find the basic data. Cans which exhibit so-called 'breaks' in heating curves do not completely obey either heat transfer model, nor do those showing irregular shaped cooling curves. Further, the pure conduction model is limited in its application because assumptions that the can contents are homogeneous, and that thermal properties are not temperature dependent, are almost invariably made to simplify the mathematics.

2.4. The Overall Method

The overall method to determine the lethality of a process with respect to either pathogens or spoilage organisms is therefore

(1) Determine the temperature/time profiles at points in the can as a function of heating time (t_h), retort temperature (θ_h), cooling

time (t_c) and cooling water temperature (θ_c). Direct measurement or prediction by a heat transfer model may be used. The latter still requires measured data.

(2) Use either eqn (9) or eqn (11) to determine rate constants so that a rate constant/time relationship can be established at each point where the temperature profiles were determined.

(3) Determine values of C_f, the survivor concentration at these points. Equation (7) yields

$$C_f = C_0 \exp\left[- \int_0^{t_f} k \, dt \right]$$ (15)

The integration may be carried out numerically or analytically. Analytical integration is only possible if temperature/time histories were determined from one of the pure mode heat transfer models.

(4) Evaluate N_f. Using eqn (1) gives

$$N_f = \int_0^{V_c} C_f \, dV$$ (16)

This integration may also be carried out either analytically or numerically. Analytical integration is again restricted to cases where either the pure convection or pure conduction model is used to determine the temperature/time histories throughout the can.

(5) Calculate an F value using a rearrangement of eqn (4):

$$F = D \log (N_0/N_f)$$ (17)

If there is a uniform distribution of spores in the can prior to processing then

$$N_0 = C_0 V_c$$ (18)

In practice it would be unusual for this full method to be applied as it has been described here. This is because researchers have been able to provide calculation methods and data sheets that simplify the procedures. The thermal processing literature contains a number of approaches for carrying out the general method described in Steps (1) to (5). These approaches have used different simplifying assumptions, and hence there are different limitations on the final results obtained from them. All approaches require the use of data that can only be determined experimentally so the usefulness of any calculated results is further limited by data uncertainties.

Inspection of eqn (17) shows that a small percentage uncertainty in F leads to a large percentage uncertainty in N_f. For a process with $D = 0.20$ min, F values of 5·95 and 6·00 min might appear similar. However, the N_f values are 80% apart. In assessing the accuracy of F values this factor must be considered.

Steps (1) to (5) describe the method to determine the lethality (Γ) of a process defined by θ_h, t_h, θ_c and t_c values. Often it is required to calculate the heating time t_h to achieve a certain F for set values of θ_h and θ_c. Unfortunately Steps (1) to (5) are not easy to use in reverse. Some procedures are available, and will be covered in Section 4.2. The other alternative is an iterative determination by assuming t_h values and determining F values for these.

2.5. Reasons for Underprocessing

There are three types of reasons for underprocessed cans that can be related to the lethality calculation.

(a)   Calculations of F or t_h for the process have not been sufficiently accurate because either
(i) assumptions made to derive the calculation method used are not valid for the circumstances under which it has been applied, or
(ii) the numerical work has been inaccurate (particularly in graphical integrations).

(b)   Data used in the calculations have been imprecise. These might be kinetic data for spore inactivation (D, z and E_a values), or data used to determine temperature/time profiles (α, f_h, f_c, j_h and j_c values).

(c)   Between can, and between processing run variations in practice are larger than anticipated so that the safety margin allowed is inadequate. The lethality of the least processed can is therefore below the required minimum level. Sources of problems include variations in can fill weight and hence headspace, variations in initial product temperature, differences in steam supply to different parts of a retort, differences in retort venting and come-up time from run to run and retort to retort, changes in cooling water temperature, and the possibility that F values determined in a small pilot plant retort are not directly applicable for the same t_h and θ_h in a commercial plant.

Research reported in the literature has tended to concentrate on (a)(i), but there are useful contributions in the other areas. As was mentioned in Section 1.3.1 incorrect lethality calculations have not been a major cause of botulism outbreaks.

3. ANALYSIS OF THE KINETIC MODELS USED

3.1. Determination of N_f Values
The number of surviving spores N_f (or the probability that one spore has survived) is not easily determined because different parts of the can receive different heat treatments. Hence although it is normally assumed that initial spore concentration (C_0) is uniform across a can, the concentration after processing (C_f) is position dependent. Evaluation of eqn (16) is therefore not a simple task.

3.1.1. Pure Convection
The exception is the pure convection model in which the assumed perfect mixing ensures that all the can contents are equally processed so that

$$N_f = C_f V_c \qquad (19)$$

and

$$F = D \log (C_0/C_f) \qquad (20)$$

3.1.2. Pure Conduction
The pure conduction model allows temperature/time profiles to be predicted at any point in the can so that the variation of C_f with position can be determined. Integration of eqn (16) can proceed in one of two ways. Hicks,[12] Hayakawa,[13] Teixeira et al.[14] and Thijssen et al.[15,16] all used numerical integration of eqn (16) having found C_f values at 'grid' points in the can

$$N_f = \int_0^{r_c} \int_0^H C_f \, dr \, dx \qquad (21)$$

The integration accuracy is limited only by the size of the intervals (Δx and Δr) over which the integration is carried out provided the temperature profiles were accurately determined from the pure conduction model. With modern digital computers to perform the calculations Δx and Δr can be small implying accurate integration.

The alternative approach used by Gillespy,[17] Stumbo,[3,18] Ball and Olson[19] and Jen et al.[20] was to assume that a simple mathematical relationship could be applied between lethality and position. Integration may then be a combination of analytical and numerical techniques. The accuracy of such approaches is limited by the adequacy of the form of the relationship assumed. Holdsworth and Overington[21] have questioned the accuracy of the Gillespy approach, and Hayakawa[22] discusses weaknesses of the others. In view of the increasing ease of carrying out numerical integration by computer it is unlikely that further development of such semi-analytical integration procedures is warranted.

3.1.3. All Cases
The simplest approach for integration of eqn (16) is

$$N_f \approx C_{fc} V_c \tag{22}$$

where C_{fc} is the concentration of survivors at some critical point in the can. This method can be applied whatever the heat transfer mode provided the temperature/time profile at the critical point is known. F values are often determined by applying eqn (20) at this point, although this does not match the correct definition of eqn (17). This review will refer to such a value as the 'local F'.

In order that eqn (22) will always give a safe over-estimate of N_f, the critical point should be the position in the container receiving the least heat. This is normally assumed to be the slowest heating point in the container, yet there is clear evidence that over the whole heating and cooling cycle, the least processed point may be elsewhere. Assuming that heat transfer to all surfaces of the can is equivalent, it is easily seen that for the heating phase the F_h value is at a minimum at the geometric centre. For the cooling phase alone the minimum F_c value must occur at the surface. When F_h and F_c values are summed, the position of the point with the lowest F_t value can lie anywhere between the centre and the surface depending on the relative contributions of heating and cooling. Flambert and Deltour[23] demonstrate this effect which can also be seen in work as early as the tables in Hicks' 1951 paper.[12] Holdsworth and Overington[21] list other workers who have noted the effect. The location of the least processed point lies on the vertical axis through the can for small H/r_c values, and in a ring shaped region for large H/r_c values.[23]

In spite of this evidence that C_f is higher elsewhere in the can, the slowest heating point is still commonly used as the critical point. Fortunately the difference in C_f between the least processed point and the slowest heating point appears to be small for practical process conditions (e.g. Tables 2–5 of Hicks[12]). Furthermore, C_f will be lower in the outer reaches of the can beyond the least processed point so the assumption inherent in using the slowest heating point

$$V_c C_{fc} \geqslant \int_0^{V_c} C_f \, dV \tag{23}$$

is probably valid. Equation (23) is a key assumption yet test results for a wide range of conditions have not been published.

There are some misconceptions as to how the value of C_{fc} should be interpreted. Firstly, cans with the same local F at the critical position and hence the same C_{fc} value (provided C_0 is the same) do not have the same degree of commercial sterility if they differ in size. Sterility is measured by the probability of survival of a single spore in a whole can. Therefore as can size and hence volume (V_c) increase, C_{fc} must be decreased and the local F for the process increased to achieve a constant N_f for all cans.

Secondly, an incorrect interpretation that may be made is that the numerical values of N_f and C_{fc} are approximately the same so that C_{fc} can be considered the probability of sterility for the can. If the units of spores ml^{-1} are used for C_{fc}, this implies that beyond a volume of 1 ml at the critical point the probability of spore survival is negligible. While concentrations are lower beyond the 1 ml critical region, the volume of material is large. Hicks[12] shows that for one example of a conduction heated can N_f is numerically 13 times C_{fc}. This does meet the condition of eqn (23) but shows that treating C_{fc} as equivalent to N_f is incorrect. Choice of process lethality to be achieved at a critical point (local F) should always be based on consideration of a final N_f value, not the C_{fc} value. A valuable discussion on this aspect is available.[24]

3.2. Kinetic Models for Spore Destruction

The thermal resistance of micro-organisms is an important consideration in the calculation of sterilisation processes. For thermal resistance
data to be amenable to analytical treatment there must be accepted
theories which describe the rate of death of micro-organisms. Despite
considerable investigation over the last 80 years, no one theory of
death can explain all the experimental observations.

A logarithmic decrease in the number of organisms is often
observed (Fig. 1, curve A). Two kinds of deviations from the
logarithmic order occur. The shoulder or lag (curve B) is generally
held to be a combination of some activation mechanism with
inactivation.[25] The shoulder, which is easily taken into account for the
calculation of sterilisation processes,[26,27] often becomes negligible
when destruction prevails over activation, as is the case in canning
where higher temperatures are used. It will not be considered further
here.

In contrast, the tail or tailing off of biphasic curves and the upward
concavity (curves C and D) have not gained any consensus but have
been widely reported (e.g. ref. 28). It is worth noting that the classic
work by Esty and Meyer[11] on *C. botulinum*, which forms the basis for
all low-acid canned food processes used today, showed a tail lasting
40 min. They commented that 'the mortality of the spores is a gradual
process and follows in a general way the law of logarithmic decline'.

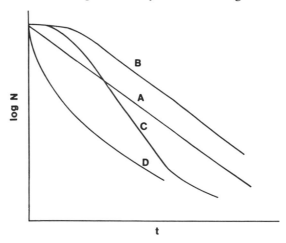

FIG. 1. Types of survivor curves. A = logarithmic; B = shoulder or lag;
C = biphasic curve with a tail; D = upward concavity.

Since it is not possible to describe curves of type C and D by a simple mathematical model, calculation of sterilisation processes for spores which exhibit such curves will be imprecise. Methods which do not rely on the shape of survivor curves have been developed. The survivor curves are plotted out as in Fig. 1 but instead of calculating a single D value, the time for a given number of decimal reductions is determined. Mossel and de Groot[29] termed this value the $MPED_n$ for 'most probable effective dose of thermal energy to achieve n overall decimal reductions'. This approach was first suggested by Ott et al.[30] and was endorsed by Moats et al.[31] and Mossel.[32] However, it may not be particularly suitable for the canning industry, since it was designed for use with measurable N_f values ($\geqslant 10^{-2}$) thus avoiding the need for extrapolation beyond the range of experiments. For practical purposes data for N_f values below 10^{-3} are not measurable so extrapolation beyond the data set, and the assumption that some sort of mathematical relationship between N and time exists, is inevitable.

If thermal death of bacteria is not truly logarithmic (and the evidence suggests that it is not), then it must be recognised that D values are of limited validity and extrapolation of D values determined from survivor curves carried through 4 or 5 log cycles to higher probabilities of kill could be seriously inaccurate. The time for a given number of decimal reductions at high N values (say $10^{10}-10^{-2}$) may not be the same as that for the same number of decimal reductions at lower N values (e.g. 10^4-10^{-8}). Therefore the use of bacteriological process evaluation (as discussed in Section 1.3.3) to confirm the calculated thermal process would seem desirable.

Procedures for mathematically analysing both logarithmic and non-logarithmic survivor curves are available.[33,34]

3.2.1. Factors Influencing the Thermal Resistance of Spores

Various factors are known to influence the thermal resistance of bacterial spores.[25] These can broadly be divided into pre-treatment influences, the conditions prevailing during the actual treatment and post-treatment recovery.[35] Of particular interest to canners is the fact that spore resistance is affected by the nature of the food including such factors as water activity, pH and the level of fats, proteins, carbohydrates and salts present.[36] Therefore, it would seem desirable, if not essential, before designing thermal processes, to have data on the thermal resistance of the spores of interest actually determined in the particular food product. Sadly, such information is very sparse. A

TABLE 2

EXPERIMENTAL VALUES OF DECIMAL REDUCTION TIMES
(MIN) FOR THERMAL DESTRUCTION OF *C. botulinum*
(62A) IN VARIOUS FOODS[37]

	$D_{115\cdot6}$ (*min*)		
pH	Spaghetti, tomato sauce and cheese	Macaroni creole	Spanish rice
4.0	0·128	0·127	0·117
4·2	0·143	0·148	0·124
4·4	0·163	0·170	0·149
4·6	0·223	0·223	0·210
4·8	0·226	0·261	0·256
5·0	0·260	0·306	0·266
6·0	0·491	0·535	0·469
7·0	0·515	0·568	0·550

Reprinted from Food Technology (1965) **19(6),** 1003.
Copyright © by Institute of Food Technologists.

good example of such data is shown in Table 2[37] which illustrates both
the effect of pH and food product on the D values for spores of C.
botulinum 62A at 115·6°C.

It has long been a generally accepted fact that a pH of 4·6 or less
will inhibit the growth of *C. botulinum* spores in food.[38] Occasionally
growth of *C. botulinum* and toxin formation at pH values lower than
4·6 have been reported. In such cases the authors have ascribed the
unexpected growth and toxin formation to local pH differences in
inhomogeneous media and growth of the organism before pH equi-
libration, or to the fact that fungi created micro-environments within
or adjacent to the mycelial mat, where the pH was higher than 4·6.[39]

However, recent work[40,41] has shown that spores of *C. botulinum*
can grow and produce toxin in aqueous suspensions of soya protein at
pH values as low as 4·2, and in skimmed milk at pH 4·4. The type of
acid present was important, hydrochloric and citric acids being much
less inhibitory than lactic and acetic acids. Clearly this is an area for
further work. To date, regulatory authorities have made no moves to
change the definition of low-acid foods (for which a full botulinum
cook is required) to include foods with a pH below 4·6.

It is worth noting that the $D_{121\cdot1}$ value of *C. botulinum* normally
used in canning calculations is based on the thermal death time data of
Esty and Meyer[11] determined in Sorenson's pH 7·0 phosphate buffer

and corrected for thermal lag by Townsend et al.[42] (The corresponding $D_{115\cdot6}$ value using $D_{121\cdot1} = 0\cdot204$ min and $z = 9\cdot78°C$ is $0\cdot755$ min). However, Esty and Meyer (ibid.) found some of their spore suspensions to be more resistant in nutrient media than in buffer. The later paper from the same laboratory[42] also cited a higher C. botulinum resistance in other heating media (peas, for example) than in phosphate buffer. They also found that the z value was consistently lower in foods than in phosphate: $7\cdot4$–$8\cdot7°C$ rather than $9\cdot1$–$10\cdot0°C$. More recently, Perkins et al.[43] have confirmed that z varies significantly from $7\cdot1$ to $8\cdot2°C$ in a range of formulated foods.

3.3. Temperature Models

Regardless of the order of death of micro-organisms, the rate constant k, for the rate of thermal inactivation of spores, is temperature dependent. Ever since the early work of Esty and Meyer,[11] it has been common practice in calculating heat sterilisation processes for canned foods to assume that the linear model could be used for temperature dependence. This was introduced earlier in eqns (10) and (11)

$$k = k_r 10^{(\theta - \theta_r)/z} \tag{11}$$

The ratio of the rate constants at two temperatures (θ_1 and θ_2) can be derived from eqn (11). Substitution of eqn (8) yields

$$D_2/D_1 = 10^{(\theta_1 - \theta_2)/z} \tag{24}$$

Applying the equivalent procedure to the Arrhenius relationship (eqn (8)) yields

$$D_2/D_1 = 10^{(1/T_2 - 1/T_1)E_a/2\cdot303R} \tag{25}$$

Equations (11) and (24) imply that z is constant over a range of temperature whereas the Arrhenius relationship by assuming that E_a is constant suggests a variable z. Figure 2 shows the difference in results arising from predictions using the two models. Over a limited temperature range where curve A is a tangent to curve B both models will satisfactorily describe the data.[17] Outside this range the curvature of the hyperbola results in the Arrhenius relationship giving (on a relative basis) conservative estimates of the efficacy of a heat process.

Judgement as to which model is the better description of measured data is difficult because experiments to measure k values have significant uncertainties associated with them, both as a result of the possible inadequacy of the first order kinetic model, and arising from

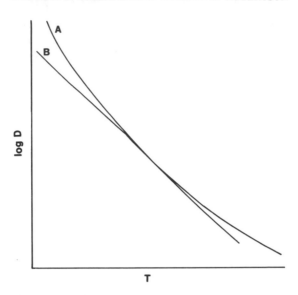

Fig. 2. Plot of log D *vs* absolute temperature T. A = Arrhenius relationship; B = linear temperature model.

the difficulties in collecting data. Some workers used only the linear form and found satisfactory fit (e.g. refs 44 and 45), whereas others (e.g. ref. 46) found that the Arrhenius relationship was satisfactory.

Jones[47] compared specimen thermal processes evaluated on the alternative assumptions of a constant z value and a constant activation energy (E_a). However, as was subsequently pointed out,[48] the comparison by Jones (ibid.) was ambiguous since he used a z value from data on *C. botulinum* in phosphate buffer and an E_a value for the same organism heated in pea puree. The z values from these two sources differ so much that calculations on this basis alone would not yield comparable results.

This objection was overcome with the work of Jonsson *et al.*[49] who compared both models over the temperature range 111–125°C using spores of *Bacillus stearothermophilus* heated in neutral phosphate buffer. They found that while both temperature models gave very good prediction of the shape of the relationship in the measured data, they had a significant lack of fit. Since both models had a low predictive capacity, they suggested that sterilisation calculations should be based on experimental results obtained at the actual temperatures to be used, rather than on extrapolations of data from other temperatures. They also questioned whether it is correct to regard spore heat inactivation as a simple first order reaction in view of the significant lack of fit of the Arrhenius model.

3.3.1. Application of Temperature Models to Square Processes

A 'square' process is one in which the contents of a container are lifted uniformly and instantaneously to the process temperature at time zero, and cooled uniformly and instantaneously at the end of the heating period. The necessary process time to achieve a required F can be calculated algebraically. Ball and Olson[19] present the appropriate equation for the linear form

$$F_r = t_t 10^{(T - T_r)/z} \tag{26}$$

$$= t_t 10^{(\theta - \theta_r)/z} \tag{27}$$

where F_r is the number of equivalent minutes at temperature θ_r. The corresponding formula using the Arrhenius relationship is

$$F_r = t_t \exp\left[-(E_a/R)(1/T - 1/T_r)\right] \tag{28}$$

The discrepancy between the two methods in predicting the processing time can be calculated by taking ratios of t_t values calculated from eqns (26) and (27). Defining t_{tl} as the time calculated by the linear form, and t_{ta} as that by the Arrhenius relationship, then

$$t_{tl}/t_{ta} = \exp\left[2 \cdot 303(T_r - T)/z - (E_a/R)(1/T - 1/T_r)\right] \tag{29}$$

Using the data of Jonsson et al.[49] for B. stearothermophilus ($z = 8 \cdot 5°C$, $E_a = 343 \cdot 5$ kJ mol^{-1} in the range 111–125°C) the data in Table 3 were calculated. The reference temperature should always be taken in the

TABLE 3
RELATIVE STERILISATION TIMES (t_{tl}/t_{ta}) FOR SQUARE PROCESSES CALCULATED FROM EQN (29).
$z = 8 \cdot 5°C$; $E_a = 343 \cdot 5$ kJ mol^{-1}

θ	Values of θ_r (°C)			
	111·1	116·1	121·1	126·1
101·1	1·176	1·208	1·197	1·147
111·1	1·000	1·026	1·017	0·975
121·1	0·983	1·009	1·000	0·958
131·1	1·106	1·135	1·125	1·078
141·1	1·410	1·447	1·434	1·374

temperature range studied. The results show that the two methods for calculating sterilising times are approximately equivalent for the temperature range at which the data were determined, but differ significantly beyond this.

Over the temperature interval 101·1°C to 131·1°C, the method commonly used by the canning industry gives sterilisation times which are up to 20% longer than the method based on the Arrhenius model for a reference temperature of 121·1°C. Table 3 has been included to show how much the uncertainty in data can affect results. On the basis of available data neither model can be considered as better than the other in spite of the discrepancies.

3.3.2. Application of Temperature Models to Non-Square Processes
The term 'non-square' can be applied to any process in which temperature changes with time. The local F value can be determined using

$$F_r = \int_0^{t_h} L \, dt \qquad (30)$$

where L values are calculated using eqns (9) and (12) in the case of the Arrhenius relationship, and eqns (11) and (12) in the case of the linear form. The results have been designated F_{ra} and F_{rl}, respectively. Table 4 gives data for a process in which temperature changes with time using the same *B. stearothermophilus* data as Table 3. Again differences result from the use of the two models, the size of the difference being increased by wider separation of θ_h and θ_r.

TABLE 4

RELATIVE LETHALITY VALUES (F_{ra}/F_{rl}) FOR PROCESSES CHARACTERISED BY $f_c = f_h = 20$ min; $j_c = j_h = 1·0$; $t_h = 30$ min; $\theta_{in} = 40°C$; $\theta_c = 20°C$; $z = 8·5°C$; $E_a = 343·5$ kJ mol^{-1}

θ_h	Values of θ_r (°C)			
	111·1	116·1	121·1	126·1
101·1	1·323	1·358	1·346	1·290
111·1	1·062	1·090	1·080	1·036
121·1	0·979	1·005	0·996	1·001
131·1	1·025	1·053	1·044	1·001
141·1	1·212	1·244	1·234	1·182

3.3.3. Summary

Although calculated results suggest that the two kinetic models are not equivalent, measured data are sufficiently imprecise that neither model can be regarded as a better fit of the data. Values of z and E_a for the same organism vary between foods which also increases the uncertainty in calculations for temperature effects. Unless more precise data can be measured that show that one model is a better description than the other, the continuing use of both can be justified. It appears that the choice between them by researchers probably reflects no more than a belief that one or the other is more mechanistically correct or easier to use.

4. ANALYSIS OF HEAT TRANSFER MODELS

When analysing heat transfer during thermal processing, the aim is to determine as accurately as possible temperature/time profiles at points of interest in the container. There may be only a single critical point if eqn (22) is used to evaluate N_f, or a number, should more sophisticated evaluation of N_f via eqn (16) or (21) be required. The temperature/time profiles may be obtained by direct measurement, or by a combination of theoretically derived heat transfer equations and measured data.

4.1. Direct Measurement

Practical considerations in placing thermocouples or other temperature measuring devices limit direct measurement to only a few positions in the can. Hence direct measurement is almost invariably used in conjunction with the single critical point approach for N_f. The position at which measurements are taken is usually the geometric centre because the true 'least-processed' point cannot be easily determined. Apart from the factors discussed in Section 3.1, the least processed position is often offset from the can geometric centre by non-symmetrical external heat transfer (a useful analysis for one example case is given by Naveh et al.[50]), complex heat transfer in the can headspace,[51] and possibly by mixing effects that occur as the result of boiling and condensation in the can. It has not been proven that the condition in eqn (23) is always met for practical situations, yet there is no practical alternative to using the can geometric centre.

The aim in measuring temperature/time profiles is to establish the degree of variation around the average so that assessment of the worst case can be made. Rather than compare temperature/time profiles, it is normal to compare F values derived from the so-called 'General Method'. This method in its simplest form involves the following calculation

$$F = \sum_{i=0}^{i=I} (L_i \, \Delta t) \tag{31}$$

where

$$L_i = 10^{(\theta_i - \theta_r)/z} \tag{32}$$

The process is divided into I intervals of Δt such that

$$I \, \Delta t = t_t \tag{33}$$

Using the experimental temperature/time data the mean temperature θ_i for each time interval Δt is found and L_i determined using eqn (32). Although the numerical integration in this manner is conceptually simple, a variety of tables and special graph papers exist to simplify the calculations. Other numerical integration procedures can be used (e.g. Patashnik[52]). Provided the combination of integration procedure and choice of Δt does not contribute significantly to the overall uncertainty in F, the selection of procedure is unimportant. As shown later, other uncertainties are substantial.

Direct measurement of F values by a biological indicator unit has been demonstrated by Pflug et al.[53] and Navankasattusas and Lund[54] used electrical analogues to measure accomplished lethality. However, the General Method is most often applied by use of thermocouples to measure temperatures, and subsequent numerical or graphical evaluation of eqns (31), (32) and (33).

If a number of experimental investigations for the same product, the same type of can, and the same nominal process are undertaken, a range of F values will result. In commercial practice it is normal to test several cans in the same retort run, and perhaps perform several runs. Hence between run and between can (within a run) variations will be evident.[55]

For a safe process the worst case can should be used in calculations. Under circumstances in which the contribution of the cooling phase to F_t is ignored, Hicks[56] stated that the worst case can is the slowest heating can in the test. The standard reference book for canners in the USA[9] suggests one test run of six to eight cans may provide sufficient data, but that many runs may be necessary to ensure that the conditions of slowest heating have been included in the study. Lund[57] states that the slowest heating container in a sample of 20 is common commercial practice in the USA. The UK guide[58] suggests that 'a minimum of three separate heat penetration runs should be carried out with a minimum of three replicate cans in each run to obtain results in close agreement before a provisional scheduled heat process based on the lowest value obtained is set.' Codex Alimentarius[2] states 'it is essential to carry out an adequate number of heat penetration tests to determine the variations which should be taken into account in the scheduled process.'

The alternative approach in assessing the worst case is by use of statistics to estimate the likelihood of any cans having an F below a certain value. This approach removes doubt as to whether the slowest heating container was included in the study, but relies on knowing what statistical distribution the results follow. Robertson and Miller[55] showed that there was no consensus in the literature on the distribution. It is likely that the form of distribution depends markedly on the experimental techniques of the investigator.

4.1.1. Effect of Complex Heat Transfer

Although pure conduction and pure convection heat transfer models are often used in studies on canning, there are a wide variety of situations in canning where neither model is appropriate. In particular, the cooling phase is often characterised by irregularly shaped temperature/time profiles. These curves are considered to arise from boiling and condensation in the can, and from mixing of the can contents. Board et al.[10] studied these effects in a systematic manner for baked beans which heated by conduction. They found that variations in can headspace, seaming pressure, air pressure in the retort during cooling and the temperature of the can contents at the end of the heating phase significantly affected F_c. Their study was limited by the lack of an accurate method for predicting F_c values that would have arisen if conduction was the sole means of heat transfer in the can during cooling. Cleland and Gesterkamp[59] used finite difference predictions of temperature profiles plus numerical integration of L values to obtain accurate F_c values for pure conduction cooling. Their measured F_c values were as little as 19% of those expected for pure conduction, but the percentage changed markedly between 19 and 100% if the four factors discussed above were varied. Both papers therefore show that there is inherent danger in expecting F_c to be the same for all cans unless headspace, can seaming pressure, retort pressure during cooling and product temperature at the end of heating are closely controlled.

Jones et al.[60] demonstrated the effect of fill weight (and hence headspace) on measured F values in cans of green beans. A physical model to describe how boiling and condensation led to mixing of the can contents was partly verified by Cleland and Gesterkamp.[59]

The uncertainty in F_c resulting from complex heat transfer has led to recommendations that the required F_t be achieved in the heating phase and the cooling phase be regarded as a safety factor.[10] This approach, while safe microbiologically, may have detrimental effects on product quality.

The implication of complex heat transfer on measured F values is that imprecise control of the four factors outlined above will increase the variation of F_t between cans, and between retort batches. As well as the effect on F_c, some effect on F_h has been noted.[59] The assessment of what is the worst case F_t for process design will be of lower accuracy if complex heat transfer occurs.

4.1.2. Temperature Measurement Errors

Common problems in temperature measurement are poor calibration of equipment, conduction along thermocouple leads altering the temperature of the measurement junction, and imprecise placement of the thermocouple junction.

A systematic 1°C error in measured temperatures corresponds to a 26% inaccuracy in F for $z = 10°C$.[12]

Conduction along the leads has most effect on measured temperature profiles early in both the heating and cooling phases.[61,62] The F value is more affected by the errors at the onset of cooling (where it is underestimated) than by temperature errors in heating (where it is overestimated). Errors reduce with reduction in wire diameter, but quantitative guidelines are difficult to establish because of the large number of factors affecting heat transfer to the measurement probe. If the contribution of the cooling phase is taken into account, F_t is unlikely to be overestimated (due to underestimation of F_c), but should the cooling phase be ignored and $F_t = F_h$ assumed, then overestimation of lethality can occur. Practical considerations prevent the use of very fine wire[55] so some inaccuracy in a measured temperature profile, and hence F, due to heat conduction along thermocouple wires is inevitable.

Imprecise thermocouple placement will lead to F variation from can to can, and so is assessed by using a number of cans.

4.1.3. Can to Can Variations Within a Retort Batch

Variations can arise from the factors outlined in Section 4.1.1 (complex heat transfer and variation in thermocouple placement). They are also contributed to by variation in the product between cans if it is not homogeneous. A further factor is can placement within the retort. There may be variations in external heat transfer to the cans, when cooling is started there may be several minutes from when the first can is immersed until the last one is covered, and the water temperature may vary with position in the retort. The test runs carried out should encompass variation of all these factors if a reasonable estimate of F for the worst case is to be obtained.

While the procedure of taking the lowest F in a small number of cans (discussed in Section 4.1) is simple, it may not always give a reliable estimate of a typical worst case in commercial operation where the numbers of cans are very large. The alternative of assuming that a certain statistical distribution can be applied is also not exact as the appropriate distribution is often not known. To illustrate the two approaches, data for seven cans in the same retort batch were taken from Table 2 in Robertson and Miller.[55] The F_o values were 25·3, 24·3, 26·6, 25·9, 28·1, 27·3 and 28·3 min. The sample of seven cans met the suggested USA guideline of six to eight cans.[9] For the illustration it was assumed that the data were normally distributed, and the Student's t-distribution was consulted to establish the F_o that 99% of cans would be expected to exceed. The value obtained was 21·9 min yet the lowest value in the sample was 24·3 min. For small samples better estimates of extreme cases are likely to be obtained from a statistical approach than simply choosing the worst case in the sample. The percentage difference between 21·9 and 24·3 min may appear small. However, substitution of these values in eqn (17) would lead to estimates of N_f several log cycles apart.

4.1.4. Variations Between Retort Batches

Processes of the same nominal length at the same steam temperature may not be equivalent. Factors that could lead to F variation are differences in retort loading, poor steam temperature control, inadequate venting for some batches, poor control of presssure during cooling, and differences in the length of the retort come-up period. Variations between batches can be assessed either statistically or by choosing the worst case run. The data in Table 2 of Robertson and Miller[55] cover eight runs each of the same seven cans. The lowest value of F_o was 21·0 min. An example of a feasible statistical procedure is to treat all 56 cans as one normally distributed population. The F_o value that 99% of all cans would be expected to exceed was 20·5 min. The worst measured case and a statistical estimate would be expected to agree better as the number of data are increased.

4.1.5. Scale-up to Commercial Retorts
The lethality achieved for a given steam temperature and heating time combination in a pilot plant retort may not be representative of what would occur in commercial retorts. Differences in retort come-up time and in can stacking (which affects heat transfer from the steam to the can) are important. Verification of process lethality in commercial retorts is therefore recommended.

4.1.6. Use of the General Method
Section 4.1 has shown the development of the General Method. This method covers all five steps introduced in Section 2.4. As was pointed out in Section 3.1.3, if a local F value is to be used for the critical point, the chosen F should reflect the desired N_f value for the whole can. Although in the General Method the two integrations (across the can volume, and across time) are performed numerically, the major sources of uncertainty in values of N_f may lie in other areas such as

(a)   the validity of using first order kinetics;
(b)   the uncertainty in choice of temperature model, and in data used in the models such as z and E_a;
(c)   can to can variations;
(d)   batch to batch variations and scale-up and
(e)   complex heat transfer.

Cannery technologists and regulatory authorities must accept the available techniques and data for (a) and (b), but make their own judgement on the other three aspects. Neglect of the cooling phase provides a safe manner of treating (e), but this may be at the expense of some product quality loss.

Published guidelines for handling (c) and (d) are vague. The least processed can from a small number of cans and a small number of runs may not yield a good estimate of the worst case in commercial operation. However, a statistical approach will not necessarily lead to better information as the distribution that results follow appears to vary. Some workers[55,63] found that a normal distribution was a good description of their data, yet Powers *et al.*[64] collected data following a skewed distribution.

A useful statistical analysis on uncertainties was carried out by Lund[57] but commercial canners' guides[9,58] still prefer the approach of taking the worst can from a small sample. The success of the latter approach in commercial practice probably reflects the large but ill-defined safety factors in most thermal processes.[65] A need for further experimental work and statistical interpretation of this is indicated. This research should have the aim of establishing more quantitative guidelines for cannery technologists and regulatory authorities (who have to approve processes) on how to decide the number of cans and number of retort runs to investigate, and how to interpret the resulting data. Problems in the application of the General Method are most likely to be in this area.

4.2. Formula Methods as Data Extenders

The General Method for lethality evaluation provides a means of evaluating F for different sets of processing conditions under which experimental measurements were made, but does not allow the investigator to predict, without further experimentation, F values for different conditions, or conditions that will achieve a required F. For the latter two situations a group of methods known as Formula Methods have been developed. The Formula Methods do not replace data collection. The data they require differ in form from those needed for the General Method; instead of the full temperature/time history only four parameters are required. These are f_h, a parameter related to the slope of a plot of $\ln(\theta_h - \theta_{fc})$ *vs* time, j_h, the dimensionless intercept of such a plot at time zero, f_c, a parameter related to the slope of a plot of $\ln(\theta_{fc} - \theta_c)$ *vs* time, and j_c, the dimensionless intercept of such a plot at 'steam-off' time. An alternative approach where heat transfer is totally by conduction uses the thermal diffusivity α instead of f and j values. Analytical expressions exist to relate α to f and j.

4.2.1. Range of Application

Formula Methods cannot be applied across the full spectrum of lethality calculations. In fact, their application is restricted to the following four situations.

(A) Contribution of the cooling phase ignored and a single critical point approach is used to determine N_f. Heat transfer need not be pure conduction or pure convection. Formula methods can be applied provided the graph of $\ln(\theta_h - \theta_{fc})$ *vs* time is straight (non-linearity can be tolerated when θ_{fc} is low). There may be more than one straight line segment provided the break occurs at the same temperature in all cans. The value of N_f is accurate provided eqn (23) holds.

(B) Contribution of the cooling phase is ignored and N_f obtained by integration across the can volume. Heat transfer within the can must be by pure conduction or convection. There can be only one linear segment in the $\ln(\theta_h - \theta_{fc})$ *vs* time graph. More than one indicates complex heat transfer which prevents prediction of temperature profiles in different parts of the can.

(C) F determined over the full process. A single critical point is used to evaluate N_f. The measured temperature/time profile at the critical point must plausibly fit a pure conduction model although heat transfer need not be purely by this mode. This is because the temperature/time profiles used by most Formula Methods have been chosen to fit the conduction case. Equation (23) is assumed to hold. More than one linear segment in the heating curve can exist provided the break occurs at the same temperature in all cans.

(D) F evaluated over the full process. N_f is evaluated by integration across the can volume. Heat transfer within the can throughout heating and cooling must be by pure conduction or convection.

In reality Formula Methods in categories (B), (C) and (D) are limited almost completely to internal heat transfer by conduction because the pure convection model assumes no temperature difference across the can contents, which is impractical for all but thin liquids.

4.2.2. Can to Can and Batch to Batch Variations

As is the case with the General Method, users of Formula Methods wish to assess the lethality in the least processed can. The values of f_h, j_h, f_c, j_c and α must therefore reflect the behaviour of this can. Hence these data should be determined for a number of cans and a number of batches. Variations will arise from the same sources as were important for the General Method. These have been discussed in Sections 4.1.2, 4.1.3 and 4.1.4. Few guidelines exist for choosing appropriate f, j and α values from replicate experiments. Hicks[56] suggests that the can with the highest f_h should be used. This is appropriate for methods in categories (A) and (B) but may not suit (C) and (D). There appears to be no standard statistical procedure for choosing appropriate data.

Inaccuracy in application of Formula Methods can arise from inadequacies in the methods themselves, from inaccuracy in the way in which the f, j and α values represent a typical limiting case can, and from violation of assumed heat transfer modes. The first factor has been researched extensively, but the usefulness of such research can be questioned if the latter two aspects are more limiting in practice. A search of the literature failed to reveal any quantitative study of the relative contribution of the three types of imprecision.

If even a small percentage of test cans show irregular cooling curves it is extremely unwise to use Formula Methods in categories (C) and (D).

4.2.3. Problems in Derivation of Formula Methods

A search of the literature reveals a number of different Formula Methods. These have arisen because the mathematics involved in Steps (1), (2) and (4) of Section 2.4 are too complex for analytical solution without simplifying assumptions. The assumptions vary depending in which of the four categories a method lies. Methods in categories (C) and (D) can usually be applied to situations where the cooling phase is ignored, in which case any assumptions relating to the cooling phase are not limiting.

Inherent in all Formula Methods is the assumption that a plot of $\ln (\theta_h - \theta_{fc})$ vs time has a straight line segment. Hence all Formula Methods make the following assumptions

(a) θ_h remains constant throughout the heating phase.
(b) The thermal properties of the can contents are not temperature-variant.
(c) The retort come-up time is negligibly small.

Although these are the mathematical conditions for a linear plot, plausible straight lines will result in practice if violations are not large. Particularly important in this respect is (c). A straight line plot does not develop for some time in a conduction process so significant deviations from a square process can occur yet still yield a straight line. As j_h is affected significantly by (c) a Formula Method should not be applied routinely to a process with a significant come-up time unless this come-up time is repeatable from batch to batch.

A Formula Method that is designed to determine N_f by integration across the can volume has further limitations

(d)   The product is homogeneous.
(e)   The can is assumed to have no headspace.
(f)   External heat transfer to the can is even on all surfaces.
(g)   All resistance to heat transfer is internal to the can contents. Throughout the heating phase the surface temperature of the can contents is the steam temperature, and during cooling it is the (constant) cooling water temperature.

Formula methods covering the cooling phase further assume

(h)   There is an instantaneous change from heating to cooling.

Although these assumptions appear to greatly restrict their usefulness, Formula Methods do provide the only way of predicting F for sets of conditions that differ from those under which experiments have been conducted, and also allow prediction of process times to achieve a required F value.

4.2.4. Review of Formula Methods for a Single Critical Point

Category (A) situations are covered by all Formula Methods. The appropriate equations were first developed by Ball.[66] If, as is usually the case, spore destruction prior to the onset of the exponential heating curve is insignificant then

$$F_h = f_h \exp\left[(\theta_h - \theta_r)/z\right](E_1[(\theta_h - \theta_g)/z] - E_1[(\theta_h - \theta_{in})/z])/2 \cdot 303 \quad (34)$$

$$\theta_g = \theta_h - j_h(\theta_h - \theta_{in})10^{(-t_h/f_h)} \quad (35)$$

This formula is mathematically exact. Differences between F values from it and the General Method arise only from imprecise data. Examples are included by Board et al.[10] and Robertson and Miller[55] which show correspondence of F values within a few per cent.

Direct back-calculation to t_h values to achieve a required F_h is not possible. A value for θ_g must be found by trial and error solution of eqns (34) and (35), or published tables of data can be interpolated.

The extension of a Formula Method to take into account the cooling phase (Category C) was first attempted by Ball.[66] The main problem in trying to develop a mathematical formula for the cooling phase is that most of the spore destruction that can be attributed to cooling occurs early in the cooling cycle during which time a number of terms in the infinite mathematical series defining the temperature/time profiles are significant. The main limitation of Formula Methods covering cooling arises from the type of approximation used for the temperature profiles over this critical period.

Merson *et al.*[67] have provided a theoretical analysis of Formula Methods, whereas Smith and Tung[68] assessed numerically the prediction accuracy of various Formula Methods applied to pure conduction situations. In the latter study the benchmark for comparison of the Formula Methods was a finite difference heat transfer calculation and a numerical integration to obtain lethalities. This approach predicts closely lethalities obtained from eqns (34) and (35). The finite difference method uses the same equations for the heating and cooling phase so therefore accurately predicts temperatures for the cooling phase. Disagreement of F_t values between the finite difference method and a particular Formula Method most likely indicates that the Formula Method does not accurately predict the actual cooling phase lethality for the conduction case. A similar analysis to that of Smith and Tung[68] was made by Naveh *et al.*[69] They used both experimental data and finite element predictions, but concentrated their study on the cooling phase.

It was convenient in the present study to categorise the methods in a similar manner to that used by Smith and Tung.[68] The organisation of the methods in this way is consistent with the historical development of Formula Methods in the literature.

(a) *Ball's method.* An arbitrary decision that a hyperbola could be used for the temperature profile in the early part of cooling was made. It was assumed that $j_c = 1.41$ and $f_h = f_c$. Numerical results were tabulated.[66] Some mistakes in these tables have been reported.[68,70,71] The method underpredicts F_t by substantial amounts for the conditions of the Smith and Tung survey. The F value for a process for which $(\theta_h - \theta_g) = 0.5°C$ was underpredicted by about 15%.[68] The corresponding N_f values would be several log cycles in error, fortunately on the conservative side. The assumptions of a hyperbola and $j_c = 1.41$ therefore introduce significant inaccuracy to the formula.

The assumption that $f_h = f_c$ is often criticised. If heat transfer is by pure conduction, and if the external heat transfer coefficient is large, then f_h and f_c are equal. Significant differences in the two f values may indicate complex heat transfer, in which case it is probably inadvisable to apply a Formula Method at all (due to can to can variations in the nature of complex heat transfer[59]). The condition that $f_c = f_h$ is therefore not as restricting as it may appear. In their book Ball and Olson[19] proposed a new Formula Method taking into account variation in j_c. The approximation for the temperature profile in the early stages of cooling was still a hyperbola which would limit the accuracy of the method. Hicks[72] and Steele et al.[73] have published corrections to the tables in the book. Shiga[74] suggests a variant that allows easy interpolation of data.

(b) *Gillespy's method*. The approach used by Gillespy[17] was to assume that the contribution of the cooling phase to the total lethality was the same as if the heating phase had been extended by $0.03f_h$ min. The factor of 0.03 was arrived at by an examination of experimental data. Smith and Tung did not include this method in their study but it would not be expected to be accurate across a wide range of conditions.

(c) *Stumbo's method*. Stumbo and Longley[75] extended the range of Ball's method by publishing tables for various j_c values. In his book Stumbo[3] revises these tables using temperatures calculated from finite difference heat transfer calculations followed by numerical integration to obtain lethalities. Hence these tables assume pure conduction and make no arbitrary assumption about the shape of the cooling phase temperature profile. However, Stumbo used a short-cut procedure to obtain data for a range of j_c values. Rather than vary combinations of t_h and θ_h to obtain data for different j_c at the centre position, he took data from other parts of the can to get different j_c values. A full j_c range cannot be covered at the centre position, but probably the most important parts of the range could have been. The method is limited by this aspect, and by rounding errors in the numerical calculations. The disagreement of F values with the finite difference results of Smith and Tung (up to 20% but usually less than 5%) probably reflects these problems. Of the Category (C) methods tested by Smith and Tung, this was most accurate yet it still predicted N_f values that could be several log cycles in error, again on the conservative side. Tung and Garland[76] have developed computer programs to implement the procedure.

(d) *Hayakawa's method*. Hayakawa[77] approached the problem of accurately obtaining temperature profiles for the early part of cooling in a different way that followed in part earlier work by Jakobsen.[78] Hayakawa curve-fitted experimental data with a set of relatively simple mathematical equations. The tables derived were subsequently corrected and extended.[79] They cover a range of j_c and do not require that $f_c = f_h$. Their use in relating t_h and F_t relies on the curve-fitted relationships being accurate. If $f_c \neq f_h$ complex heat transfer that varies between cans could be the cause. In such a case the tables may not give good predictions. For the conduction situation studied by Smith and Tung, the performance of this method was poorer than Stumbo's but better than Ball's.

(e) *Steele and Board's method*. Griffin *et al.*[80] proposed a different mathematical equation to approximate the temperature profile at the onset of cooling to those used by Ball[66] or Hayakawa.[77] The temperature difference between the actual product temperature and that predicted by extrapolating the logarithmic part of the cooling curve was assumed to decrease logarithmically with time. There is no allowance for the temperature rise in the early stages of cooling. Steele and Board[81] adopted this approximation for the temperature profiles but introduced a new and useful concept known as a sterilising ratio to simplify the resulting tables which were calculated by numerical integration of the equations used. The calculations carried out by Smith and Tung suggest that this approach led to only slightly better approximation of temperature profiles in the conduction situation than the hyperbola used by Ball.[66] For pure conduction the Stumbo approach was consistently better.

(f) *Other approaches*. A number of other approaches for total lethality determination at a critical point have been proposed but not converted to convenient tables or alignment charts for routine use across a wide range of conditions. These have used accurate analytical series solutions to predict temperature/time profiles for heating and cooling by conduction, and are potentially at least as accurate as the Stumbo method. Important contributions of this nature are those of Hicks,[12] Flambert and Deltour[82,83] and Lenz and Lund.[84]

Clark[85] described a general numerical approach to lethality determination, and Skinner[86] proposed a linear dynamic model suitable for use where external conditions change.

4.2.5. Formula Methods for Mass Average Survivor Determination
Category (B) and (D) methods are of necessity restricted to situations where heat transfer through both the heating and cooling phases is by conduction. (It should be noted that determination of mass average

survivor levels is a trivial extension of Category (A) and (C) methods for the pure convection case.) Heat conduction formulae are used to determine temperature/time profiles at different points in the can, survivor concentrations at these points calculated, and an integration across the can volume carried out to estimate N_f. The method of integration has been discussed in Section 3.1.

Holdsworth and Overington[21] and Hayakawa[22] present valuable analyses of Formula Methods of this type. The important methods are as follows.

(a) *Gillespy's method*. The assumptions made[17] were that the cooling phase lethality was equivalent to an extra $0.03f_h$ min of heating, and that there was a linear relationship between the local F for any 'iso-F' shell in the can, and the volume enclosed by this shell. The latter assumption cannot be reconciled with results such as those of Hicks[12] which indicate that the lowest lethality is not necessarily at the can centre. The method is therefore not expected to be accurate across a wide range of conditions.

(b) *Stumbo's method*. The method is described in Stumbo's book.[3] The assumption about a linear relationship between local F for a particular iso-F shell, and the volume enclosed by the shell, resembles that of Gillespy. Using an example of a No. 10 can, Stumbo[18] showed that the relationship holds reasonably well for the innermost 19% of the can volume. He argued that inaccuracy can be tolerated in regions beyond this as the outer regions contribute little to N_f when eqn (16) is integrated. A remarkably simple formula for the mean F value resulted. As the method was derived for a specific can size it would seem advisable to check results from it across a range of conditions before adopting it for general use. Jen *et al.*[20] published tables to allow the extension of the method to a wide range of z values.

(c) *Ball and Olson's method*. In their book Ball and Olson[19] divide the can into 11 pseudo-isothermal regions and compute C_f values for each at the end of the process. Values of j_c for the outer parts of the can are low so sudden drops in food temperature at the onset of cooling were assumed. In practice drops of this nature do not occur except at the can outer surface. Test results verifying the accuracy of the method over a wide range of conditions have not been published. Many of the weaknesses in Ball's earlier work for a single critical point have not been corrected in the mass average method so it would be surprising if the method was accurate across a wide range of conditions.

(d) *Analytical heat conduction calculations, numerical integration over volume.* Hicks,[12] Thijssen *et al.*,[15,16] Hayakawa[87] and Lenz and Lund[88] used this approach. Provided all non-negligible terms in the Fourier series are used, the temperature/time profiles at any point in the can will be accurately determined. Numerical integration of eqn (16) yielded N_f values. Hayakawa[89] presents computer programs to carry out calculations, whereas Lenz and Lund[88] give examples of alignment charts that can be used for cans with $H/2r_c = 1\cdot12$. It is unclear from the literature how widely these methods are used. They are expected to be accurate for situations where heat conduction is the only heat transfer mode.

(e) *Numerical heat conduction calculations, numerical integration over volume.* It is well established that numerical techniques such as finite differences closely model heat conduction. Thus results using finite differences[14,23] should give virtually identical results to methods covered under (d), and hence are accurate. The computer programs are much simpler than those using the analytical solutions as many terms are needed in the latter at the onset of cooling.

There appears to be a need to verify results from the Stumbo and Ball and Olson formulae against results from either group (d) or (e) if the former are to be used routinely with confidence.

4.2.6. Other Aspects of Formula Methods
A number of papers have been published to help canners prepare data for use in Formula Methods, and to provide guidelines to take account of retort come-up time. In the introduction to their paper, Uno and Hayakawa[90] provide a useful bibliography for the latter aspect.

5. DISCUSSION AND CONCLUSIONS

A number of inadequacies in the heat transfer and kinetic models used for thermal processing have been highlighted in this review, yet the long and continuing success of commercial canners in producing adequately processed products must also be acknowledged. The large safety factors inherent in most commercial processes would seem to explain why this is so.

Future research will be most valuable if it is directed towards areas that allow canners to minimise safety margins and consequently reduce both quality degradation and costs. Such areas are summarised below.

(a) *Inadequacy of the kinetic models*. If, in spite of substantial conflicting data, first order kinetics and a constant z value continue to be used, values of C_f and N_f predicted from temperature/time histories will be subject to doubt. There seems to be no practical alternative to using the current models, but it is advisable to choose target N_f values that include a margin for deviation from logarithmic kill. Useful research might be directed towards establishing quantitative guidelines for safety margin selection.

(b) *Imprecision of kinetic data*. D and z values vary substantially between strains of the same organism, and with the composition of the food. More data collection will not necessarily lead to 'better' values to use. Instead, research could establish quantitatively a degree of variation in z that should be included in the process design calculations, and improve approximate procedures for quickly estimating the effect on F (and hence N_f) of changing z.

(c) *General Method*. This method is more limited by the accuracy of the measured temperature/time profiles than by the mathematics used. Research could establish better guidelines for determining how to take batch to batch and can to can variations into account. Experimental investigation of a variety of statistical approaches may be warranted.

(d) *Formula Methods for a single critical point*. Caution must be exercised in applying a Formula Method to any can that exhibits irregular heating or cooling curves that cannot be satisfactorily resolved as straight line plots on logarithmic temperature difference vs time plots. Quantitative evaluation of methods in the manner of Smith and Tung[68] is valuable. There appears to be a need for the f_h/F vs $(\theta_h - \theta_g)$ tables of Stumbo[3] to be updated so that as far as is possible data for different j_c values are established at the geometric centre of the can. For the heating phase all Formula Methods are mathematically accurate. In view of Smith and Tung's results, when the lethality of the whole process is considered methods other than that of Stumbo should be used with caution.

(e) *Formula Methods for mass average survivor levels.* Although there has been a considerable amount of research, methods of this type have not been widely used by canners. There appears to be a need for a study of the type carried out by Smith and Tung[68] to provide guidance. Sufficient doubts exist so that the Gillespy, Ball and Olson and Stumbo approaches cannot be recommended until verified by such a study. Presentation of the results from methods categorised in Section 4.2.5 under (d) and (e) in a form that is both easily used, and covers a wide range of conditions, would be desirable.

(f) *Data for Formula Methods.* Development of techniques to help the canner decide what values of f and j should be taken from a set of measured data to represent the worst case has been limited. Further research, perhaps incorporating a statistical approach, is warranted.

(g) *Application to computers.* The digital computer has been used by researchers for lethality calculations, particularly those involving the Formula Methods. A number of programs have been published. A commercial software package, available both on floppy disk for micro-computer, and on magnetic tape for larger computers, could find a ready market. Such a package should encompass user-friendly features in input/output and carry out calculations by all important lethality estimation procedures. An international working party to define the basis of the package could be established, to provide overall guidance to the software developer. Publication of listings of programs in the literature can no longer be regarded as an effective way of making programs available to canners. Programs such as those of Smith and Tung[68] or Hayakawa[89] could provide a partial basis of a package if the normal level of commercial input/output was added and the range of methods expanded. Finite difference calculations should play a major role in predicting lethality for pure conduction cases without resort to f_h/F vs $(\theta_h - \theta_g)$ tables.

In the foreseeable future market demands are likely to encourage cannery technologists to adopt processes that minimise product quality degradation. This will require reductions in safety margins and therefore techniques for assessment of what is a safe process must be refined. This review has attempted to make cannery technologists and regulatory authorities more aware of the limitations of the techniques currently in use, and to suggest subjects that future research might address to augment the current process design procedures.

REFERENCES

1. SMELT, J. P. P. M. and MOSSEL, D. A. A., In *Principles and Practice of Disinfection, Preservation and Sterilisation*, 1982, A. D. Russell, W. B. Hugo and G. A. J. Ayliffe (ed.), Blackwell Scientific Publications, Oxford.
2. Codex Alimentarius Commission, *Recommended International Code of Practice for Low-Acid and Acidified Low-Acid Canned Foods*, 1979, Vol. G—Ed. 1; Codex Alimentarius Commission. Joint FAO/WHO Food Standards Programme.
3. STUMBO, C. R., *Thermobacteriology in Food Processing* (2nd edn), 1973, Academic Press, New York.
4. STUMBO, C. R., PUROHIT, K. S. and RAMAKRISHNAN, T. V., *J. Food Sci.*, 1975, **40**, 1316.
5. PFLUG, I. J. and ODLAUG, T. E., *Food Technol.*, 1978, **32(6)**, 63.
6. ALSTRAND, D. V. and ECKLUND, O. F., *Food Technol.*, 1952, **6**, 185.
7. WHITTER, L. D., In *Food Microbiology*, 1983, A. H. Rose (ed); (Economic Microbiology, vol. 8), Academic Press, London.
8. YAWGER, E. S., *Food Technol.*, 1978, **32(6)**, 59.
9. National Canners Association, *Laboratory Manual for Food Canners and Processors*, 1968, Vol. I, Avi, Westport, Connecticut.
10. BOARD, P. W., COWELL, N. D. and HICKS, E. W., *Food Research*, 1960, **25**, 449.
11. ESTY, J. R. and MEYER, K. F., *J. Infect. Dis.*, 1922, **31**, 650.
12. HICKS, E. W., *Food Technol.*, 1951, **5**, 134.
13. HAYAKAWA, K., *Food Technol.* 1967, **21**, 1073.
14. TEIXEIRA, A. A., DIXON, J. R., ZAHRADNIK, J. W. and ZINSMEISTER, G. E., *Food Technol.*, 1969, **23(3)**, 352.
15. THIJSSEN, H. A. C., KERKHOF, P. J. A. M. and LIEFKENS, A. A. A., *J. Food Sci.*, 1978, **43**, 1096.
16. THIJSSEN, H. A. C. and KOCHEN, L. H. P. J. M., *J. Food Sci.* 1980, **45**, 1267.
17. GILLESPY, T. G., *J. Sci. Food Agric.*, 1951, **2**, 107.
18. STUMBO, C. R., *Food Technol.*, 1953, **7**, 309.
19. BALL, C. O. and OLSON, F. C. W., *Sterilisation in Food Technology* (1st ed.), 1957, McGraw–Hill, New York.
20. JEN, Y., MANSON, J. E., STUMBO, C. R. and ZAHRADNIK, J. W., *J. Food Sci.*, 1971, **36**, 692.
21. HOLDSWORTH, S. D. and OVERINGTON, W. J. G., *Calculation of the Sterilising Value of Food Canning Processes*. Technical Bulletin No. 28, 1975, Campden Food Preservation Research Association, Chipping Campden, Gloucestershire, England.
22. HAYAKAWA, K., *Food Technol.*, 1978, **32(3)**, 59.
23. FLAMBERT, C. M. F. and DELTOUR, J., *Lebens.-Wiss. u. Technol.*, 1972, **5(1)**, 7.
24. HICKS, E. W., *Food Technol.*, 1952, **6**, 175.
25. RUSSELL, A. D., *The Destruction of Bacterial Spores*, 1982, Academic Press, London.

26. KING, A. D., BAYNE, H. G. and ALDERTON, G., *Appl. Environ. Micro.*, 1979, **37**, 596.
27. PFLUG, I. J., In *Industrial Sterilisation*, G. B. Phillips and W. S. Miller (ed.), 1973, Duke University Press, North Carolina.
28. CERF, O. *J. appl. Bact.*, 1977, **42**, 1.
29. MOSSEL, D. A. A. and DE GROOT, A. P., In *Radiation Preservation of Foods*, Publication 1273, 1963, pp. 233–264, Washington National Academy of Science.
30. OTT, T. M., EL-BISI, H. M. and ESSELEN, W. B., *J. Food Sci.*, 1961, **26**, 1.
31. MOATS, W. A., DABBAH, R. and EDWARDS, V. M., *J. Food Sci.*, 1971, **36**, 523.
32. MOSSEL, D. A. A., *Microbiology of Foods*. 3rd edn, 1980, Utrecht: The University of Utrecht, Faculty of Veterinary Medicine.
33. HAYAKAWA, K., *Can. Inst. Food Sci. Technol. J.*, 1982, **15**, 116.
34. HAYAKAWA, K., MATSUDA, N., KOMAKI, K. and MATSUNAWA, K., *Lebens.-Wiss. u. Technol.*, 1981, **14**, 70.
35. MOLIN, G., In *Principles and Practice of Disinfection, Preservation and Sterilisation*, 1982, A. D. Russell, W. B. Hugo and G. A. J. Ayliffe (ed), Blackwell Scientific Publications, Oxford.
36. HANSEN, N. H. and RIEMANN, H., *J. appl. Bact.*, 1963, **26**, 314.
37. XEZONES, H. and HUTCHINGS, I. J., *Food Technol.*, 1965, **19**, 1003.
38. ITO, K. and CHEN, J. K., *Food Technol.*, 1978, **32(6)**, 71.
39. ODLAUG, T. E. and PFLUG, I. J., *Appl. environ. Microbiol.*, 1979, **37**, 496.
40. RAATJEES, G. J. M. and SMELT, J. P. P. M., *Nature*, 1979, **281**, 398.
41. SMELT, J. P. P. M., RAATJEES, G. J. M., CROWTHER, J. S. and VERRIPS, C. T., *J. appl. Bact.*, 1982, **52**, 75.
42. TOWNSEND, C. T., ESTY, J. R. and BASELT, F. C., *Food Res.*, 1938, **3**, 323.
43. PERKINS, W. E., ASHTON, D. H. and EVANCHO, G. M., *J. Food Sci.*, 1975, **40**, 1189.
44. RAHN, O., *Bacteriol. Rev.*, 1945, **9**, 1.
45. STUMBO, C. R., MURPHY, J. R. and COCHRAN, J., *Food Technol.*, 1950, **4**, 321.
46. WANG, D. I., SCHARER, J. and HUMPHREY, A. E., *Appl. Micro.*, 1964, **12**, 451.
47. JONES, M. C., *J. Food Technol.*, 1968, **3**, 31.
48. COWELL, N. D., *J. Food Technol.*, 1968, **3**, 303.
49. JONSSON, U., SNYGG, B. G., HARNULV, B. G. and ZACHRISSON, T., *J. Food Sci.*, 1977, **42**, 1251.
50. NAVEH, D., PFLUG, I. J. and KOPELMAN, I. J., *J. Food Sci.*, 1984, **49**, 461.
51. EVANS, H. L. and BOARD, P. W., *Food Technol.*, 1954, **8**, 254.
52. PATASHNIK, M., *Food Technol.*, 1953, **7(1)**, 1.
53. PFLUG, I. J., JONES, A. T. and BLANCHETT, R., *J. Food Sci.*, 1980, **45**, 941.
54. NAVANKASATTUSAS, S. and LUND, D. B., *Food Technol.*, 1978, **32(3)**, 79.

55. ROBERTSON, G. L. and MILLER, S. L., *J. Food Technol.*, 1984, **19**, 623.
56. HICKS, E. W., *Food Res.,* 1961, **26**, 218.
57. LUND, D. B., *Food Technol.*, 1978, **32(3)**, 76.
58. ANON, *Guidelines for the Establishment of Scheduled Heat Processes for Low-Acid Canned Foods. Technical Manual No. 3.*, 1977, Campden Food Preservation Research Association, Chipping Campden, Gloucestershire, England.
59. CLELAND, A. C. and GESTERKAMP, M. F., *J. Food Technol.*, 1983, **18**, 411.
60. JONES, A. T., PFLUG, I. J. and BLANCHETT, R., *J. Food Sci.*, 1980, **45**, 217.
61. COWELL, N. D., EVANS, H. L., HICKS, E. W. and MELLOR, J. D., *Food Technol.*, 1959, **13**, 425.
62. BEVERLOO, W. A. and WELDRING, J. A. G., *Lebens.-Wiss u. Technol.*, 1969, **2**, 9.
63. TOEPFER, E. W., REYNOLDS, H., GILPIN, G. L. and TAUBE, K., *US Department of Agricultural Technology Bulletin No. 930*, 1946, p. 28.
64. POWERS, J. J., PRATT, D. E., CARMON, J. L., SOMAATMADJA, D. and FORTSON, J. C., *Food Technol.*, 1962, **16**, 80.
65. BOARD, P. W., *C.S.I.R.O. Division of Food Research Circular No. 7*, (2nd edn), 1977.
66. BALL, C. O., *Thermal Process Time for Canned Food. Bulletin 7–1(37)*, 1923, National Research Council, Washington DC.
67. MERSON, R. L., SINGH, R. P. and CARROAD, P. A., *Food Technol.*, 1978, **32(3)**, 66.
68. SMITH, T. and TUNG, M. A., *J. Food Sci.*, 1982, **47**, 626.
69. NAVEH, D., PFLUG, I. J. and KOPELMAN, I. J., *J. Food Process. and Preserv.*, 1983, **7**, 275.
70. FLAMBERT, C. M. F., DELTOUR, J., DICKERSON, R. W. and HAYAKAWA, K., *J. Food Sci.*, 1977, **42**, 545.
71. STEELE, R. J. and BOARD, P. W., *J. Food Sci.*, 1979, **44**, 292.
72. HICKS, E. W., *Food Res.,* 1958, **23**, 396.
73. STEELE, R. J., BOARD, P. W., BEST, D. J. and WILLCOX, M. E., *J. Food Sci.*, 1979, **44**, 954.
74. SHIGA, I., *J. Food Sci.*, 1976, **41**, 461.
75. STUMBO, C. R. and LONGLEY, R. E., *Food Technol.*, 1966, **20**, 341.
76. TUNG, M. A. and GARLAND, T. D., *J. Food Sci.*, 1978, **43**, 365.
77. HAYAKAWA, K., *Food Technol.*, 1970, **24**, 1407.
78. JAKOBSEN, F., *Food Res.*, 1954, **19**, 66.
79. DOWNES, T. W. and HAYAKAWA, K., *Lebens.-Wiss. u. Technol.*, 1977, **10**, 256.
80. GRIFFIN, R. C. JR., HERNDON, D. H. and BALL, C. O., *Food Technol.*, 1971, **25**, 36.
81. STEELE, R. J. and BOARD, P. W., *J. Food Technol.*, 1979, **14**, 227.
82. FLAMBERT, C. M. F. and DELTOUR, J., *Lebens.-Wiss u. Technol.*, 1972, **5(2)**, 72.
83. FLAMBERT, C. M. F. and DELTOUR, J., *Industr. alim. agr.*, 1973, **90(1)**, 5.
84. LENZ, M. K. and LUND, D. B., *J. Food Sci.*, 1977, **42**, 989.

85. CLARK, J. P., *Food Technol.*, 1978, **32(3)**, 73.
86. SKINNER, R. H., In *Control of Food Quality and Food Analysis*, 1983, G. C. Birch and K. J. Parker (ed.), Elsevier Applied Science, London.
87. HAYAKAWA, K., *Can. Inst. Food Sci. Technol. J.*, 1969, **2**, 165.
88. LENZ, M. K. and LUND, D. B., *J. Food Sci.*, 1977, **42**, 997.
89. HAYAKAWA, K., *Adv. Food Res.*, 1977, **23**, 75.
90. UNO, J. and HAYAKAWA, K., *J. Food Sci.*, 1981, **46**, 1484.

Chapter 2

REFRIGERATED STORAGE OF PACKAGED MEAT

H.-J. S. NIELSEN

Food Technology Laboratory, Technical University, Lyngby, Denmark

SUMMARY

A vast amount of species of micro-organisms are present on meat at time of packaging. The micro-organisms capable of development in packaged meat are, however, more limited, and highly influenced by intrinsic and extrinsic factors. The general picture is one of a change from a dominance of catalase positive to catalase negative bacteria, regardless whether the meat is fresh or cured and/or cooked. In both fresh and cured and cooked meat products, temperature highly influences microbial growth. Permeability of packaging film and gas atmosphere favour lactic acid bacteria and help controlling other organisms. The amount of glucose in and the pH of fresh meat and the curing ingredients in processed meats are important factors for microbiological development.

The initial composition of the flora may disturb the flora change during storage, and thus high levels of bacteria like B. thermosphacta, Enterobacteriaceae *or* Vibrio *may positively compete with the lactics, and dominate, dependent on product type and storage conditions. Pathogenic bacteria will not normally present a problem in packaged meat, growth being controlled by the competing flora, especially the lactic acid bacteria, in combination with the different processing factors. Accidental contamination with higher levels of pathogens in an otherwise nearly sterile product, may develop to levels normally considered hazardous to health. Low temperature storage will, however, arrest proliferation of most of the pathogens in such cases.*

45

1. INTRODUCTION

Economically, packaged meat is one of the most important groups of food today. This demands that great efforts are made so as to market meat which is not spoiled due to development of micro-organisms. Different factors are involved in controlling growth of bacteria in packaged meat. Some of these apply only to cured, cooked meat products, others also to packed fresh meat. Many studies have been done to elucidate the influence of the different preservative factors, both on the saprophytic flora and on pathogenic bacteria. In meat, where the levels of the preservative factors are within the range acceptable for consumption, growth will inevitably occur sooner or later. The use of the different preservative factors, however, act differently on the many species initially present on the meat, and thus allow development of the least unacceptable, the saprophytic bacteria. Furthermore, the factors should be of a nature which does not allow growth of pathogenic bacteria.

2. SOURCES OF CONTAMINATION

The interior of the flesh of animals slaughtered according to normal hygienic procedures is usually sterile.[1] Therefore contamination during slaughtering and processing makes up the flora in meat products together with micro-organisms from other ingredients. In raw cured meats like Wiltshire bacon, contamination with micro-organisms takes place by brine injection and eventually tank curing.[2] Thus it has been shown that bacteria (atypical streptobacteria) may follow the meat during processing and end up as important spoilage organisms in the finished bacon.[3] In products where spices and herbs are added, these may greatly contaminate the products in spite of the low concentrations used. Commercial natural spices may have microbial counts of up to 10^8/g, with a high proportion made up of sporogenic bacteria,[4] and herbs may also be heavily contaminated. These problems may, however, be reduced by different kinds of decontamination procedures like ionization[5] or gassing with ethylene oxide;[6] the latter procedure has been prohibited in several countries.

In cooked meat products, the heat treatment greatly influences the survival of micro-organisms in the finished product. Warnecke *et al.*[7] showed that the numbers of micro-organisms surviving the heat treatment (to a centre temperature of 68°C) was unaffected by the bacterial load of the raw sausage mix, while the counts in finished frankfurters were related to the quality of the raw mix.[8] The sensory qualities of the finished products were, however, in both studies highly correlated with the microbial quality of the raw products. Studies of the heating process of Bologna-type sausage showed that most of the Gram negative flora was reduced when temperatures of 56°C were reached and virtually eliminated at 67°C (with one exception).[9] At 67°C, most of the Gram positive flora was eliminated as well. The micrococci, however, and the sporogenic flora survived. At temperatures of 76–77°C, only the spore-formers survived, the flora being reduced from 10^{8-9}/g to less than 10^3/g.[10] Likewise the bacterial flora was reduced by $7 \log_{10}$ cycles when heating frankfurters to 73·9°C.[11] Similar results have been obtained in other studies.[4,12,13] *Micrococcus*, lactics and *Enterococcus* may occasionally survive, dependent on the heat treatment.[13–15]

Pasteurisation times of luncheon meat, converted to minutes at 70°C (P_{70}), showed that *Lactobacillus*, *Brochothrix thermosphacta* and *Micrococcus* dominated the surviving flora at $P_{70} = 40$, while a more severe heat treatment ($P_{70} = 105$) eliminated all but *Bacillus* and *Micrococcus*.[16]

In pasteurised meat products, the microbial quality of the raw meat and the ingredients especially influences the concentration of spore-forming bacteria in the end product, while most of the vegetative flora is eliminated during heat treatment. In raw cured meat products, on the other hand, the initial bacterial load will be influenced during processing and by contamination from the brine, together with surface contamination during handling. If the processing includes smoking, this may reduce the numbers of micro-organisms,[2,17] and especially favour the lactic flora.

The microbial quality of the freshly pasteurised meat product is good with respect to content of the vegetative flora. Therefore, the initial flora in the meat product must be related to contamination during handling and slicing from workers and slicing and packaging equipment. With different groups of indicator bacteria, it was shown that these could be isolated from workers' hands during production and on the slicing and packaging line.[18] Also lactics have been isolated from equipment and on the sliced and packaged products in spite of a total elimination of these bacteria during pasteurisation.[12] Other studies on luncheon meat products have shown similar results.[18–20]

Cross-contamination between the raw and the pasteurised product results in a product with a high microbial load in spite of adequate heat treatment.[21]

The microbial contamination of the finished meat product is the result of a number of factors. In non-heated products, the microbial contamination of the raw meat influences the microbial quality of the finished product together with micro-organisms added during processing. In heated products the main factor of contamination is the handling of the product after pasteurisation.

3. POST-PACKAGING PASTEURISED PRODUCTS

As pointed out above, pasteurisation of the meat products greatly reduces the bacterial content, but contamination takes place again during further handling of the product. Pasteurisation in a plastic bag, with subsequent storage of the bag, therefore should prolong the storage period (shelf life). In accordance with this, Schmidt[22] observed that a heat treatment of frankfurters at 80°C for 20 min and subsequent storage at 18–20°C results in a storage time three times that of a product without pasteurisation. The procedure was not possible with sliced ham because of the cook-out of fat. The effect of post-pasteurisation of various cooked cured meat products at 72°C and 80°C for 5 or 15 min showed that already a treatment of 5 min at 72°C considerably reduced the bacterial load of the product, and at a subsequent storage at 8°C for 3 weeks, the differences between the treated and the non-treated products were often several $\log_{10}$ units.[23] No growth was observed during 6 weeks storage at 10°C of luncheon meat sausage cooked in a PVDC bag to a centre temperature of 70–75°C, while a further 16 weeks storage resulted in a bacterial increase of approx. $4 \log_{10}$ units.[15] *Lactobacilli* grew at the surfaces of the sausages, while at the centre *Enterobacter* spp. dominated. At ambient temperature (25°C), abundant growth of *Bacillus* was observed during the first week, but was subsequently replaced by *Streptococcus* spp. It was suggested that *Streptococcus* grew only after the nitrite concentration was reduced due to reduction by the *Bacillus* spp.

A more thorough investigation of the influence of pasteurisation time on the microflora developing during subsequent storage was done by expressing the time–temperature sequence of minutes at 70°C

(P_{70}).[16] At a P_{70} of 40–90, spoilage associated with *Streptococcus* occurred at 25°C, at P_{70} 105–120 *Bacillus* was followed by *Streptococcus* and at even higher heat treatment (P_{70} 135 and above) spoilage was due to *Bacillus* only. The outgrowth of *Bacillus* at higher heat treatments only was apparently due to the absence of heat shock at the short heat treatments. A pasteurisation equivalent to $P_{70} = 75$ was considered adequate for the product.

The domination of different micro-organisms at the various heat treatments results in different spoilage patterns. Low heat treatments or underprocessing may result in growth of several members of *Lactobacillus* and *Enterococcus* resulting in a decrease of pH;[16,24] enterococci (*Streptococcus faecalis* subsp. *liquefaciens*) may break down gelatine, due to gelatinase activity.[25] In packages with spoilage due to *Bacillus/Streptococcus* or *Bacillus* alone a higher pH and surface softening and gas production is observed.[16] The pH being highest in packages with only *Bacillus* starch breakdown by the *Bacillus* spp., with subsequent release of glucose[26] and utilisation of this by streptococci, was suggested.[16] In underprocessed meat products or packages with leakage different spoilage patterns may arise. Especially Gram negative bacteria with low heat resistance (*Enterobacteriaceae*, *Pseudomonadaceae*, *Alcaligenes*) may be isolated.[27] Greening of whole sausages may occur by cutting due to highly contaminated ingredients and low heat treatment. This greening may be associated with several lactic acid bacteria.[28,29]

4. MICRO-ORGANISMS IN MEAT

Microbiological studies of packages taken at occasional stages during storage may show very different levels of less than $10^3/g$ to more than $10^7/g$.[30–33] High numbers may be the results of adverse storage conditions or heavy initial contamination[34] or low levels of curing ingredients. High levels may, however, not necessarily mean that the product is organoleptic unacceptable. This depends on the type of bacteria and the temperature during storage, i.e. the extent of microbial activity. The species and levels of bacteria vary considerably; this may be the result of factors treated in other sections as well as the use of appropriate selective media. Bacteria often isolated from meat include *Micrococcaceae*, lactics and *B. thermosphacta*. Several studies have shown that the species isolated often do not correlate with reference strains[35,36] and the 'atypical streptobacteria' often constitute $\frac{1}{2}-\frac{3}{4}$ of the total lactic flora in both bacon and cured and cooked products.[37-40] Studies have shown that the streptobacteria termed atypical may, however, form distinct groups.[41] Many Gram negative bacteria are isolated from meat and meat products. *Pseudomonadaceae, Moraxella, Acinetobacter, Enterobacteriaceae* and *Vibrio* are frequent isolates. Studies have also shown that a large part of the Gram negative bacteria are untypeable, and are termed atypical *Vibrio, Moraxella*-like, etc. Nevertheless many of these bacteria were shown to have distinct biochemical patterns.[36,39-42]

Generally the isolation rate of pathogenic bacteria in packaged meat is low. Sporeforming pathogens have, however, been found in vacuum-packed bacon for instance,[44] and also bacteria like *Y. enterocolitica,*[45,46] *Salmonellae*[47] and *Aeromonas hydrophila*[48] have been isolated.[47] Isolation rates are higher on fresh than on processed meat products.[45,47]

5. PACKAGED FRESH MEAT

The normal flora of packaged meat initially consists of many different species: Gram positive bacteria including micrococci, *B. thermosphacta* and lactic acid bacteria, and Gram negative bacteria like *Pseudomonas, Acinetobacter, Moraxella* and *Enterobacteriaceae* and several others, in various studies.[49-52] During early storage, growth of *B. thermosphacta,* lactic acid bacteria and Gram negative bacteria occur.[49-54] Although generally a change in flora from catalase positive to catalase negative bacteria occurs during storage in vacuum packages, the aerobic Gram negative bacteria *Pseudomonas, Moraxella* and *Moraxella*-like bacteria develop either only during the first 2–4 weeks storage $(1-2°C)$[54,56,57] or throughout the storage period.[50,53,55,58] Anyway, these bacteria remain at high levels during vacuum packaged storage.[56,59,60] Growth of *Enterobacteriaceae* may be profound during storage, however, often with a lag phase of 1–1·5 weeks,[50,54,59] a large proportion made up of *Klebsiallae.*[55,61] The Gram positive flora during storage largely consists of *B. thermosphacta,* which may show profound growth, and lactics, which generally eventually dominate in vacuum packages,[55,58] although studies on lamb have shown that *B. thermosphacta* may constitute the highest proportion of the flora during the whole storage period.[50,62] Other micro-organisms may proliferate during vacuum packaged storage, *Corynebacteria, Enterococci* yeasts, however, mostly constituting a minor proportion of the total flora.

6. PACKAGED CURED/COOKED MEAT PRODUCTS

Generally the spoilage pattern is rather similar in cooked cured meat products. Many studies have shown that while at the beginning of the storage period the flora is dominated by a catalase positive bacteria, during storage there is a shift towards a dominance of catalase negative bacteria. The initial flora may consist of micrococci, *B. thermosphacta* and Gram negative bacteria, while the levels of lactic acid bacteria and yeast normally are small.[64,65,66] During the first few weeks storage at 5–8°C, the numbers of *B. thermosphacta* increase quite rapidly, and from initial counts of less than $10^2/g$, numbers of $10^{5-7}/g$ may be reached. At an even higher temperature (10°C) these counts may be reached within 1·5 weeks. Higher initial counts result in a still faster development with millions/g of *B. thermosphacta* within 1·5 weeks.[33,67] At high initial numbers *B. thermosphacta* normally can compete quite well with other bacteria, but often when only low numbers are present initially, a decline is observed after the first few weeks increase.[65–68]

Although the lactics initially often constitute a very small proportion of the total flora ($<10–10^2/g$), these bacteria normally dominate the flora during storage, regardless of temperature. At low temperature (2°C), this may not happen until after 2–4 weeks, at 8–10°C after 2–3 weeks and at even higher temperature (20°C), within a week.[12,40,69–71] While the micrococci may constitute a high proportion of the initial flora they cannot compete with *B. thermosphacta* and lactic acid bacteria (unless the salt concentration is high), even if the increase may be quite high, to $10^{6-7}/g$.[40,64,71]

While the presence of *Enterobacteriaceae* has been low in some studies[34,72] these bacteria may, however, show considerable increases in vacuum-packed cured cooked meat products. At high temperatures (20–25°C), growth to approx. $10^{7-8}/g$ may happen within a few days storage;[40,70,71] at 10–12°C these numbers may be reached after 1–1·5 weeks. At refrigeration temperatures (2–5°C), growth to high numbers may occur after 2–3 weeks storage, but growth may be absent, controlled by a concomitant flora or factors like salt and nitrite. Other

Gram negative bacteria which may show considerable growth during storage are *Vibrio*.[73,74] High numbers of these bacteria (10^7/g) may develop at both low and high storage temperatures. Also strictly aerobic bacteria, *Moraxella/Moraxella*-like bacteria, may develop in commercially vacuum-packed cured meat products,[74] although the influence on the organoleptic properties is probably small. Especially at high temperature, growth of a number of Gram negative bacteria including *Vibrio*, *Proteus*, *Enterobacter* and *Hafnia* may be profound in packaged bacon leading to a putrefactive spoilage pattern.[75,76] *Proteus* especially were shown to be important in bacon at 22°C.[76]

Several studies have focused on yeast, showing that even at low temperature (5°C) with initial counts below 10^2/g, numbers may increase to 10^{4-5}/g within 2–3 weeks storage.[14,33,71] On account of the large size of yeast cells, numbers of this magnitude may be important because of biochemical activity.

7. PATHOGENS IN PACKAGED MEAT, TEMPERATURE EFFECT

Normally the level of contamination of pathogens in packaged fresh or cured meats is low. However, storage conditions and/or processing factors may determine the possible proliferation of these bacteria. Storage temperature is maybe the most important factor controlling development of pathogens. Normally it will act in combination with other factors like sodium chloride and nitrite, as well as packaging conditions and the presence of a competing flora. However, even if these factors fail, keeping low storage temperatures would render the meat free from growth of a number of pathogens which may be present in the package.

Staphylococcus aureus is one of the bacteria highly influenced by storage temperature within the range found in refrigeration cabinets.[77] At low temperatures of about 1·7–5°C, several studies have shown that no growth will occur in fresh or cured meat,[78–81] not even in the absence of a competing flora. Only one study has shown some increase in counts after storage at 5°C.[82] At 8°C, growth to high levels in pure culture on vacuum-packed cured meat was reported,[81] and also, in mixed culture studies at 8°C and 10°C, small but significant increases in numbers have been observed,[81] although in other studies of fresh or cured meat at 10°C, no growth of *S. aureus* was observed together with a competing flora.[85] At even higher temperatures (12–15°C), growth is rapid in combination with a competing flora[41,87] on cured meat products. However, studies on fresh meat showed that severe temperature abuse was necessary for growth of *S. aureus*. At the low pH on fresh meat (pH 5·5), growth was negative at 20°C and at 30°C growth happened only aerobically.[84] Similar results were obtained on vacuum-packed cured meats; on low pH products, 30 days storage at 4°C followed by 24 h at 30°C did not result in growth of *S. aureus*, while on a product with higher pH, these conditions gave profound growth.[85]

Studies on *B. cereus* and *C. perfringens* have not shown growth at 10°C or less with a competing flora.[79,80,85] *B. cereus* grew at 10°C in pure culture and at 12°C with a mixed flora,[81] while *C. perfringens* did not multiply at 10°C or 12°C either alone or with a mixed flora.[81] Neither was growth of *C. perfringens* observed at temperatures up to 25°C in packaged bacon.[44] Studies with *C. botulinum* (A, B, E) showed that no growth or toxin was produced at 5–7°C.[86,87] At 10°C, *C. botulinum* type E produced toxin only in ox tongue and at 100 spores/g, but not in other products.[86] At 15°C and higher toxin was also produced by type A.[83]

The lower limit for growth of salmonellae is about 7–8°C.[88] At 8°C growth was reported in a Bologna product produced without nitrite using high inoculation levels, while no growth was observed in normal sausage.[89,90] Growth of *S. typhimurium* was reported at 10°C in fresh beef, when packaged in oxygen permeable films but not in vacuum packages,[91] at 15°C, rapid growth was also observed in vacuum,[81] while *S. enteritidis* grew also at 10°C in pure and mixed culture. At temperature abused conditions salmonellae multiplied rapidly whether the packages were fresh or previously stored at refrigeration temperatures.[84,85,92]

Aerobic growth of *Y. enterocolitica* may be rapid even at a temperature of 0°C or a few degrees above that on meat.[93] Anaerobically, growth was negative even at 20°C and 30°C on fresh meat when present together with the normal flora,[84] while growing in pure culture. In vacuum-packed fresh meat, *Y. enterocolitica* grew with the normal flora at 2·5°C, although much slower than in aerobic packages,[94] the differences between packaging materials being much smaller in pure culture on packaged cured cooked meat.[81] *Y. enterocolitica* was inhibited in vacuum packages by a lactic flora at low but not at high temperature, while growth was inhibited both at 5°C and 8°C in normal contaminated packages.[81]

In conclusion it seems that generally growth of pathogens is highly dependent on storage temperature. At temperatures of 2–5°C most pathogens are inhibited; only *Y. enterocolitica* develops at these and lower temperatures, and inhibition would have to be controlled by other factors.

8. MICROBIAL INTERACTIONS IN PACKAGED MEAT

Although accidental contamination by a specific micro-organism on an otherwise nearly sterile product may result in development in an almost pure culture, normally the initial flora consists of many different organisms. Growth during storage is not only influenced by the different preservative factors inherent in the product, but involves direct antagonism or synergism between bacteria. Many different groups of bacteria have been shown to influence each other. Growth of one organism compared to another may simply reflect differences in growth rates at the specific temperature, depletion of available essential nutrients,[97] or a different affinity for growth substrates.[98] Different substances are involved in microbial antagonism, e.g. hydrogen peroxide and specific anti-microbial substances,[96] and often organic acids (lactic acid, acetic acid) are produced parallel with inhibition caused by lactic acid bacteria.[40,99] Grau[99] showed that an amount of (undissociated) lactic acid was involved, and the effect was not merely that of reducing pH. For sorbic acid both the undissociated and the dissociated forms have been shown to be active.[100] In culture media, lactic acid production at levels of 0·5–1·0% has been shown.[101] The atypical streptobacteria were reported being poor acid producers,[101] while often found on vacuum-packed meat products and together with typical streptobacteria causing inhibition of other bacteria.[103] However, in a study involving an atypical streptobacterium inoculated in vacuum-packed Bologna-type sausage, the total lactic acid concentration (L + D lactate) was found to be 2·0–3·8 mg/g meat (0·2–0·4%) at 2°C and 4·5–7·5 mg/g (0·5–0·8%) at 8°C dependent on phosphate type and glucose addition.[102] Other studies on vacuum-packed cured cooked meat products have shown lactate concentrations of 0·7% after 8 weeks at 5°C (start value 0·25–0·45%).[12] Concentrations of 10 mg/g and 10–14 mg/g were reported after storage at 7°C and 22–25°C, respectively.[70] A concomitant increase in the proportion of lactics was observed in these studies. Concentrations of 5·0, 6·1, 6·3 and 12·2 mg/g were reported after 2 weeks storage of cured smoked and sliced pork loin at 2, 5, 10 and 20°C, respectively, also here with a dominance of lactics.[40] Under anaerobic conditions values inhibitive for growth of B. thermosphacta in broth culture were 12·6 and 15·8 mg/ml at pH 5·75 and 5·85, respectively.[104] At low, but not at high, inoculation levels 13·5 mg/ml lactate was reported inhibitive for S. liquefaciens at pH 5·5, 25°C.[99] At higher pH, growth was positive also at low inoculation level. Generally, therefore lactic acid would not as a single factor inhibit these important spoilage bacteria in vacuum-packed cooked meat products. In combination with other factors, lactic acid production could still have an effect at the increasingly anaerobically conditions within the package and contribute to the increase in the proportion of lactic acid bacteria at the expense of other bacteria.

Different lactobacilli act differently on bacteria like B. thermosphacta. It was shown that B. thermosphacta was not affected by Leuconostoc mesenteroides or L. viridescens in vacuum-packed Bologna,[68] while addition of L. brevis or L. plantarum severely restricted growth, both in culture media and on vacuum-packed Bologna.[63,68] The effect was enhanced in liquid culture at decreasing temperature and nitrite addition. A study by Collins-Thompson et al.[106] connected the late appearance of Gram negative bacteria in vacuum-packed Bologna (after 6–7 weeks at 5°C) with the depletion of glucose in the meat product. It was suggested that the lactic acid bacteria were producing an antibiotic substance dependent on glucose, and the Gram negative bacteria like Serratia could better compete with the lactics because of the utilisation of other substrates (glucose-6-phosphate). Furthermore, the production of anti-microbial substances could decrease later during storage. Other factors could, however, be involved; many lactobacilli for instance, reduce nitrite[105,107] and the Gram negative bacteria belonging to Enterobacteriaceae are known to be inhibited by nitrite.[74] This could account for a late appearance of these bacteria. Moreover, quite often the Gram negative bacteria appear rather early in the storage period[40,74] and later during storage they constitute a much smaller proportion of the total flora, as is often the situation for B. thermosphacta.[40] Factors involved could be differences in growth rates, substrate availability and the accumulation of lactic acid. pH was also suggested as part of the inhibition caused by lactics in the study by Collins-Thompson and Lopez.[68] However, also in a study concerned with pathogens, other factors than pH/lactate seemed involved. The strongest inhibition by the saprophytic flora was observed at lower temperature, but accompanied by the smallest drop in pH.[108] Direct addition of lactobacilli (mixture of *S. lactic* and *Leuconostoc citrovorum*) reduced total counts of ground beef stored in polyethylene bags at 7°C,[109] and equivalent results were obtained with muscles sprayed with lactic culture. In both situations the addition of lactics was accompanied by a drop in pH. However, in a study of beef steaks, addition of lactic culture did not improve storage life.[110] On aerobically stored meat, mixed culture studies with *Pseudomonas*, *Acinetobacter*, *Enterobacter* and *B. thermosphacta* have not shown any interaction between species as long as maximum cell numbers for one bacteria were not obtained;[111] growth rates determined which bacteria developed fastest. For aerobically stored meat, the affinity for oxygen seemed important, with *Pseudomonas* using oxygen and forcing other bacteria to obtain energy by fermentation, thus reducing growth and maximum numbers.[111] On anaerobically incubated fresh meat, *Lactobacillus* outgrew both *B. thermosphacta* and *Enterobacter*, and *Enterobacter* outgrew *B. thermosphacta*,[97] a situation which did not happen on vacuum-packed Bologna.[41] Anaerobically, *Enterobacter* apparently used glucose, the only substrate for *B. thermosphacta*, thereby inhibiting growth. Maximum numbers of *Enterobacter* inhibited growth of *B. thermosphacta* and maximum numbers of *Lactobacillus* inhibited both other bacteria, probably by producing inhibitory substances.[97] In single culture anaerobic growth would be limited by diffusion of substrate from the interior to the surface of fresh meat.[112] In mixed cultures, at substrate limitation, the affinity for the substrate seems to determine growth.[97,98]

Several studies have been concerned with the influence of lactic acid bacteria on pathogenic bacteria in packaged cured cooked meat products. In a study involving salmonella stored at 8°C for 3 weeks, addition of 10^2/g of lactobacilli completely inhibited growth of 10^4/g salmonellae on vacuum-packed Bologna.[90] The inhibition happened irrespective of nitrite addition. In packages without lactics counts of 10^7/g salmonellae were reached after 2 weeks only. Various pathogens were tested in naturally contaminated packages of Bologna,[79] and/or with a flora of lactics.[108] While *S. aureus* did not grow at 10°C in the study by Stiles *et al.*,[79] it grew at 8°C and 10°C with lactic acid bacteria, although strongly inhibited, in that of Nielsen and Zeuthen.[108] Differences in oxygen permeability in the two studies may have influenced the different results. At higher temperature, growth of *S.*

aureus was only weakly affected, maximum numbers being the same as in pure culture.[108] In naturally contaminated packages growth was not affected at 10°C or 12°C.[108] While *B. cereus* did not grow at 10°C,[79] growth happens at 12°C, however, significantly reduced by lactics.[108]

S. typhimurium could not grow at 10°C in normal contaminated packages,[108] while it competed well with lactics at 12°C.[108] The normal flora, however, showed a severe restriction of growth of both *S. typhimurium* and *S. enteritidis*.[108] Growth of *Y. enterocolitica* (serotypes 03 and 09) was strongly inhibited by lactics at 5°C and 8°C, the inhibition being less as temperature increased. Growth was also restricted by the normal flora at refrigeration temperatures.[108]

In studies with fresh meat considerably higher temperatures have been used. Pathogens were tested towards several saprophytic spoilage bacteria at 20°C and 30°C, both under aerobic and anaerobic conditions.[84] At equal initial numbers, no interaction between micro-organisms (except for *Acinetobacter*) was observed, but when low numbers (10^3/cm^2) of pathogens were tested towards high (10^{6-7}/cm^2) numbers of saprophytic organisms, several interactions were seen. Thus *S. typhimurium* and *E. coli* were inhibited by *Lactobacillus* under anaerobic conditions at 20°C, but not at 30°C, neither was growth inhibited at aerobic conditions. Bacteria inhibited by *Enterobacter* were *B. cereus*, *S. aureus* and *Y. enterocolitica* under aerobic conditions, the latter also anaerobically, where *B. cereus* and *S. aureus* did not grow, even in pure culture.

No growth of *C. perfringens* was observed, when inoculating in the interior of frankfurters, vacuum-packed with the normal surface contamination at 10°C or 12°C,[108] and neither was growth observed at 4°C or 10°C by inoculation onto normal contaminated Bologna.[79] However, as mentioned before, neither was growth observed on Bologna at 12°C in pure culture.[81] A factor which in gas packaged products act directly and is treated in that section, may act indirectly in vacuum packages and is the accumulation of CO_2 and exhaustion of oxygen by microbial activity, a factor which would only be of importance in cooked meat, where the respiration of the meat has terminated. Although the lactic acid bacteria themselves[113] are susceptible to CO_2, the gas production and oxygen depletion favour the growth of lactic acid bacteria at the expense of other bacteria.

The mutual interaction between micro-organisms is one of the many factors acting in meat products. Studies of microbiological media have shown that a number of substances may be involved. However, in studies done on meat, this has not always been possible to confirm, although the importance of lactate/pH seems fully elucidated. Other factors of importance are oxygen availability under aerobic conditions and substrate availability anaerobically. Thus *Enterobacter* may influence growth of *B. thermosphacta* and both may be affected by lactics. Results on fresh and cured cooked meat do not seem identical. Pathogens are especially affected by lactics and the whole saprophytic flora at low temperatures. At higher temperatures, inhibition may decrease, and bacteria like *S. aureus*, but also salmonellae and *Y. enterocolitica,* may develop irrespective of a concomitant flora. The sporeforming bacteria *B. cereus* and *C. perfringens* do not seem to represent a hazard in packages with a competing flora.

9. FILM PERMEABILITY

When meat, whether it is fresh or cured and cooked, is packaged in a film, the gas atmosphere around it changes during storage. Oxygen concentration is reduced and CO_2 accumulated during storage.[49,114,115] Studies with both sterile and nonsterile fresh meat and bacon have, however, shown that the change in gas composition largely is a result of the meat itself and not dependent on microbial activity.[116,117] When storing fresh meat, increase in CO_2 concentration happens within a few hours,[49] and CO_2 concentrations of 20–30% and 30–40% after 2 and 4 weeks at 0–1°C, respectively, may be observed,[54] with a concomitant decrease in O_2 content. The oxygen concentration may, however, still be about 1%. When adding curing agents, sodium chloride and nitrite, to fresh pork the respiration falls,[116] the respiration of bacon being lower than in fresh meat. In cooked meat products, where respiration by the meat has terminated, one can still observe an accumulation of CO_2. Thus CO_2 concentrations of 25% were observed in air packaged frankfurters after 2 weeks storage at 4·4°C[118] and 18% was reported after 2 months storage at 4°C of vacuum-packed frankfurters. Other products showed higher CO_2 concentrations: vacuum-packed smoked pork loins reached 95% after 6 weeks at 4°C,[119] and 60–80% were observed in vacuum-packed sliced Bologna after 5–7 weeks storage at 10°C.[40] Changes in the gas composition are dependent on gas permeability and microbial growth. The increase in CO_2 concentration is more rapid both at high and low temperature in pork packaged in a more impermeable film.[49] The same results were reported with vacuum-packed sliced bacon and cooked ham.[114] Four days storage of ham at 20°C revealed a CO_2/O_2 ratio of 80 in an impermeable film and 0·4 for polyethylene.[114]

Packaging films used in older studies had rather high gas permeabilities, and therefore the effect of vacuum packaging was rather poor.[120–122] Even in these studies, some effect could be observed; thus growth of acid producing bacteria was stimulated, and yeast and moulds inhibited by vacuum packaging. Also the importance of Gram negative bacteria fell.[64] Several studies on the effect of evacuating have been done on non-evacuated samples packed in PVC and evacuated packages in film of low permeability material, therefore not only showing the effect of evacuating but also the effect of permeability of packaging material. Such studies have shown differences of aerobic counts between PVC and vacuum packages of 2–3 log_{10} units after only one weeks storage at low temperatures (1·7°C, 2·5°C).[55,78] The lower counts obtained by vacuum packaging were the result of longer lag phase and/or lower growth rates in these packages. The same results have, however, been observed when both types of packages were of the same material with low permeability,[123] evacuation enhancing the change in gas atmosphere. *B. thermosphacta* was affected as the total aerobic bacteria, while lactobacilli were not influenced.[124]

An influence of the degree of vacuum has been reported. HI-VAC packaging (an ultra-high evacuation system) of fresh pork stored at 4–5°C showed some inhibition of total aerobic counts and *Pseudomonas* and *Enterobacteriaceae*, while lactobacilli were not affected.[125] The effect was attributed to extremely low oxygen partial pressure in the HI-VAC packages. Also studies with bacon showed a retardation of growth at a vacuum level of 64–66 mm Hg compared to 71–73 mm Hg.[126]

Studies involving packaging films with oxygen permeabilities from practically 0 to 30000–40000 ml/m² 24 h 1 atm have shown marked effect on microbial growth at −1 to 5°C. Differences in numbers of aerobic micro-organisms of 2–5 log_{10} units after only a few weeks storage at refrigerated temperatures have been reported. Aerobic bacteria show a substantial longer lag phase in low permeability film.[49,127–131] Growth rates and maximum numbers of *Pseudomonas* increase with decreasing permeability of packaging film.[128] Growth rates of *B. thermosphacta* seem less affected, while maximum numbers were considerably lower with low permeability film.[67] Further studies on fresh meat showed growth with a packaging film of a permeability of $35 \, \text{ml/m}^2$ 24 h 1 atm but not at a permeability of $1 \, \text{ml/m}^2$ 24 h 1 atm,[134] and equivalent results were obtained by Roth and Clark.[133] Generally, the development of lactobacilli is stimulated in packages of lower permeability.

While the CO_2 development was faster at higher temperature (16°C), the effect of different packaging materials on total microbial growth was less than at low temperature (2°C). However, at even lower temperature (-1 to 0°C), no effect of gas permeability on microbial growth on vacuum-packed beef was observed.[132] Interaction between permeability and temperature not only influences total bacterial growth, but also the composition of the flora.[49] Aerobically, *Pseudomonas/Achromobacter* dominated at low temperature in packaged pork, and constituted still half of the flora at high temperature (16°C), but accompanied by other Gram negative bacteria (*Enterobacter/Hafnia/Kurthia*). In gas permeable film, *Pseudomonas/Achromobacter* dominated at low but not at high temperature, where the other Gram negative bacteria and *B. thermosphacta* were more important. In gas impermeable film the proportion of *B. thermosphacta* and especially that of lactobacilli at low temperature was greater, however, the *Pseudomonas/Achromobacter* group still constituting half of the flora. At high temperature, *Enterobacter/Hafnia/Kurthia* and *B. thermosphacta* were of greater importance. These results are in accordance with studies on generation times of pure cultures. These showed that *Pseudomonas* had far the shortest generation time on aerobically stored meat at 2–15°C,[111] followed by *B. thermosphacta* and *Enterobacter, Acinetobacter* having the largest generation time. Anaerobically, generation times for *B. thermosphacta* were longer than for *Enterobacter* at 15 and 10°C while at lower temperatures (2 and 5°C) the opposite was the situation. Under anaerobic conditions, however, lactobacillus had by far the shortest generation time.[97]

A study concerning the effect of packaging film on growth and development in vacuum-packed Bologna was carried out at 2, 5 and 10°C.[135] Using permeabilities of $10–950 \, \text{ml/m}^2$ 24 h 1 atm, the greatest development of the aerobic flora, the micrococci, yeast and *B. thermosphacta* was obtained in packages of the highest permeability. The packaging films influenced growth rates and maximum counts. The opposite situation occurred for the lactic acid bacteria; lower permeability and higher temperature stimulated growth of lactic acid bacteria. Growth of *Enterobacteriaceae* was stimulated by high permeability at the lower temperatures. The other Gram negative bacteria, being typically aerobics like *Moraxella/Moraxella*-like, grew in all series, even at the lowest permeability, and profound growth of atypical *Vibrio* was also observed at all permeabilities. Growth rates and maximum counts of *B. thermosphacta* in pure culture were greater in film of higher permeability (permeabilities <1–950 ml/m^2 24 h 1 atm), the effect, however, not being so great as on fresh meat.[67] The pH was also higher in the Bologna product. Even on cured meat (corned beef) at the same pH, a greater effect has been observed (differences of 2·5 log$_{10}$ units compared with about 1 log$_{10}$ unit in the study by Nielsen,[135] between packages of 1 and about 1000 ml/m^2 24 h 1 atm.

The effect of vacuum packaging and/or permeability on pathogens has been examined by several workers. *S. typhimurium* inoculated on fresh beef remained constant in vacuum packages at 10°C, while increasing significantly in oxygen permeable packages.[91] *Y. enterocolitica* was also influenced by packaging method at 1, 2·5 and 5°C, differences between vacuum packaging and packages of PVC being up to 4 log$_{10}$ units after 3 weeks.[94] On cured meats, growth of *S. aureus* and toxin production was restricted by vacuum packaging at 10, 15 and 22°C,[83,136] and also, at higher temperature, growth was most profound in the presence of oxygen;[137] however, toxin was produced also under vacuum. At high temperatures, toxin could also be produced by *C. botulinum* (type A) in Bologna, whether the packages were evacuated or not.[83] Neither had the difference in oxygen permeability any influence on toxin formation of type A *botulinum* at 15°C in Bologna or cooked ham.[86] Further studies with Bologna have not shown any growth of *C. perfringens* at 8, 10 or 12°C irrespective of gas permeability (10–950 ml/m^2 24 h 1 atm).[81] Another spore-forming bacteria, *C. sporogenes,* was also unaffected by packaging material when grown in Bologna devoid of nitrite, while no growth was observed in its presence.[138] Availability of air was also necessary for development of *Bacillus* spp. in luncheon meat chubs stored at 25°C, growth being limited to the surface of the chubs.[139]

In studies of pure cultures of *S. aureus, B. cereus, Y. enterocolitica, S. typhimurium* and *S. enteritidis* on vacuum-packed Bologna, differences in numbers related to packaging films were usually less than 0·5 log$_{10}$ unit (in the range 0–950 ml/m^2 24 h 1 atm).[81] It has been difficult to prove the influence of permeability on growth of pathogens in pure culture, but severe restriction has been shown when a concomitant flora is present. This can probably be related to development of a flora primarily consisting of lactobacilli during storage in low permeability film.[94,135] The possible inhibitive influence of the lactic acid bacteria and the development of a gas atmosphere of high CO_2 concentration restricts growth of pathogens. However, an accidental contamination with pathogens on otherwise nearly sterile meat is not likely to be influenced by packaging conditions.

A possible explanation for the decreased growth observed in vacuum compared with aerobic growth could be the decreased oxygen concentration. However, the oxygen concentration only falls rather slowly. In pork packaged in open cans wrapped with impermeable film,[49] the concentration was about 15% after 14 days at 2°C. In bacon, whether stored in CO_2 or under vacuum, the ultimate O_2 concentration was c. 2%, while in another study it was near 0% (20°C).[114] In vacuum-packed Bologna (with a high brine concentration) the oxygen concentration never fell below a few per cent at 2 or 10°C.[40] Aerobic bacteria are able to grow at low oxygen concentrations; growth at optimal rates of *Pseudomonas* and *Achromobacter* happened down to 2% oxygen at 22°C,[140] and at 5°C generation times for *Pseudomonas* increased only below 0·8% O_2.[142] Bacteria like *E. coli* and *Acetobacter* grew at maximal rates in 1% O_2[141] and generation times for *B. thermosphacta* increased at O_2 concentrations below 0·2%.[142]

10. EFFECT OF GAS PACKAGING

The uninhibited growth of many bacteria at even low oxygen concentrations in the package suggests that the increased CO_2 concentration developing during storage is responsible for the observed inhibition. Although the effect of CO_2 increases with decreasing temperatures, as discussed below, the CO_2 production in vacuum packages is also considerably lower at the lower temperatures.[40,49]

Studies on single cultures of bacteria have verified the effect of the gas composition on growth. Experiments with spoilage organisms as *Pseudomonas* and *Acinetobacter* have shown increasing inhibition with increasing CO_2 concentrations[142-144] and the influence on pathogenic bacteria like *B. cereus*, *E. coli*, *Salmonellae*, *Y. enterocolitica* and *A. hydrophila* have not been particularly different.[113,145,146] It seems that both Gram negative and Gram positive bacteria may be inhibited by CO_2.[146-148] In a study by Gill and Tan,[146] where *B. thermosphacta* or *Enterobacter* were inhibited, 10 or 20% CO_2 did not affect *B. thermosphacta* during the first weeks in studies on beef,[148] though this was followed by a fast decrease in numbers. Two studies in broth cultures showed, however a marked inhibition by CO_2, both measured as maximum growth after a certain time[145] and as maximum growth rates.[113] There was also an effect of 100% CO_2 compared with 5% CO_2/95% N_2 or vacuum.[145] The uninhibited growth of many bacteria at low O_2 concentrations could be the reason for growth in N_2 atmospheres with a low oxygen content (0·3 or 0·5%) reported in several studies. Growth of *L. plantarum* was not inhibited by vacuum, N_2 or CO_2 compared with aerobic growth,[151] and *Lactobacillus* spp. grew at 20% and 100% CO_2 on fresh meat,[148] although somewhat inhibited by 20% CO_2 at 0°C and 100% CO_2 at both 0 and 5°C. Studies on *L. viridescens* and a species isolated from cured pork showed considerably lower maximum growth rates in 5% CO_2/95% N_2 or 100% CO_2 compared to air.[113] Although the inhibitory effect of CO_2 on the lactics was the smallest among the bacteria tested, the growth rates nevertheless also were the smallest among the bacteria tested (at 25°C).[113] Furthermore, Sutherland *et al.*[148] found that although the overall effect of CO_2 as mentioned was small, development during the first weeks at 10 and 20% CO_2 was not so fast as for other organisms. On fresh meat slices stored under hydrogen, however, generation times for lactobacilli tested were lower than for *B. thermosphacta* and *Enterobacter*,[97] at temperatures from 2–15°C. Differences may be attributed to strain differences.

Studies on fluorescent and non-fluorescent *Pseudomonas, A. putrefaciens* and *Y. enterocolitica* have shown that inhibition (measured as respiration rates) caused by CO_2 increased with increasing concentrations and reached a maximum level at about pCO_2 200 mm Hg (c. 26%).[146] *Acinetobacter*[146] and *P. fragi*[151] were increasingly inhibited by CO_2 during the whole concentration range tested. Experiments showed, however, that generation times for *Pseudomonas, Y. enterocolitica* and *A. putrefaciens* were higher at 80% CO_2 than at 20% CO_2 in air, and although inhibition of growth rates for *P. fluorescens* reached a maximum at pCO_2 about 250 mm Hg (33%), growth measured as time to reach a certain maximum number of bacteria was increasingly inhibited at increasing CO_2 concentrations up to 59%, the highest level tested.[152] The often lacking effect of growth rates, which nevertheless influences the total development, suggests an inhibitive effect during the lag phase, and thus the age of the culture is important. This has been demonstrated by Clark and Lenz,[143] who showed that applying CO_2 at 0, 24 and 48 h after inoculating meat slices had increasingly less effect on growth, while the lag phase was extended when applying CO_2 at time of inoculation only.

Temperature is important for the inhibitive effect of CO_2. Studies of *Pseudomonas fluorescens* showed that at abusive temperature, 20% CO_2 in a complex medium reduced growth rates about 20% compared with aerobic control, while the effect at refrigeration temperature (5°C) was that of an 80% reduction.[152] A linear relation between growth rate and temperature was observed. The effect of temperature has been observed in other studies of growth rates and total counts reached and on shelf life.[143,145,151] The inhibition at 18% CO_2 at 10°C was, however, not different from that at 30°C for non-fluorescent *Pseudomonas* spp.[146] In the study of Clark and Lenz,[143] 10% CO_2 gave an extended shelf life of fresh beef slices of 11, 4 and 1 days at 0, 5 and 10°C, respectively. The solubility of CO_2 increases with decreasing temperature resulting in nearly constant inhibition relative to solubility in the whole temperature range,[153] and the actual growth rate decreasing effect at increasing CO_2 concentration (0–20%) was greatest at higher temperature.[143] However, the combined effect of decreasing temperatures and increasing CO_2 concentrations gave by far the largest shelf life extension.

Experiments with *B. cereus* in solution with different bicarbonate concentrations, i.e. different pH, have shown that the relative growth rates could be related to concentration of CO_2 only, but not to those of HCO_3^- or total H_2CO_3.[154] As CO_2 concentration at pH values of one above and one below pH 6·3 is 10% and 90%, respectively, pH would seem an important component in CO_2 inhibition. However, maximum growth rates of *P. fragi* (one strain tested only) were not differently affected in liquid culture by 50% CO_2 at pH 5·7 and 6·6.[153] The situation in a package of meat with a finite CO_2 amount may, however, not be the same as in liquid culture studies with a continuous supply of CO_2.

Studies on naturally contaminated packages, with pure gases, have shown that 100% O_2 is inferior to CO_2 and N_2[51,149,155–157,158a,159] or that at least no difference between oxygen, nitrogen and air was observed.[160] When testing very pure nitrogen an effect compared with air was seen as mentioned before.[149] Even if 100% CO_2 was shown to be most effective, it has an unfavourable influence on colour, unless extremely low oxygen concentrations are found in the package. When this is taken into consideration, good results have been obtained.[161] In studies with a range of mixtures of CO_2, N_2 and O_2 it was shown that at high contamination levels the mixtures of gases were inferior to vacuum packaging, except for 20% CO_2/80% O_2, which gave equal results.[155] One hundred per cent N_2 was generally not different from vacuum at 3–7°C, while at 0°C counts were higher.[162] With lower contamination levels storage in gas atmospheres containing CO_2 gave better microbial quality than vacuum packages.[158,164,166] In other studies, few differences between gases were reported.[155,163] Generally lactics are more important in gas packages, sometimes a little later in evacuated packages. When CO_2 was present in the gas mixtures *Pseudomonas* was inhibited. *B. thermosphacta* was the major bacteria in all situations on lamb (at −1°C).[149] Strains of *Enterobacteriaceae* have been shown to occur at low oxygen and oxygen absent atmospheres.[149] The optimal CO_2 concentration for restricting bacterial growth is still uncertain. Studies on single cultures showed that the chief part of inhibition on growth rates was reached at CO_2 25–30%, above which, further effect was negligible. While total growth of single cultures was increasingly inhibited at increasing CO_2 concentrations, studies on naturally contaminated packages have been inconclusive. For maximum efficiency of gas packaging, CO_2 must be applied at time of packaging, with bacteria early in the growth phase (lag phase), and low temperatures must be applied.

Some studies have focused on a possible effect of treatment with gas mixtures on the microflora after the packages have been opened and the meat rewrapped in oxygen permeable film (PVC or other). After only a few days storage in CO_2 a mixed flora of Gram negative bacteria, lactics and others has developed. Opening the package, thereby giving access to oxygen, will stimulate *Pseudomonas* spp. which will dominate,[51] while packages with a prolonged storage in 100% CO_2, after opening retained the dominance of lactics attained during gas storage. However, the total aerobic counts, after placing CO_2 packages in air, paralleled that of initially air packaged samples. Packages stored a few days in 30% CO_2/70% N_2 retained some of the inhibitory effect after rewrapping in permeable film.[164] Pre-treating pork loins for 4 or 7 days in 50% CO_2 resulted in a lag phase of 3–4 days after rewrapping, with a subsequent bacterial increase at a rate at least the rate in oxygen treated loins. Two weeks CO_2 treatment gave an additional 3 days in the rewrapped stage, until numbers of the one week treated samples were reached.[165] Compared to continued aerobic storage, this latter procedure gave a considerably extended shelf life. Other studies have given less promising results. Pre-treatment of pork loins was not superior to vacuum packaging;[56] by weekly opening of gas packages and rewrapping, psychrotrophic counts on beef and pork and to a lesser extent lamb were numerically lower in gas treated meat, differences generally not being significant.[166]

Studies on the influence of pure gases or mixtures on cured, cooked meat products have been fewer than on fresh meat assuming the same effect. However, on bacon stored at 5°C, extended lag phases were observed—for CO_2 packages, 14 days, for vacuum packages, 7 days—compared with aerobic storage.[167] Also for frankfurters, longer lag phases were observed in CO_2 atmospheres.[168] At 4°C, counts of $10^7/g$ on frankfurters were reached after 77 days in vacuum, in N_2 $6 \times 10^4/g$ were reached after 98 days and growth stopped at $4 \times 10^2/g$ after 48 days in CO_2.[119] In pork loins, counts of $10^7/g$ were reached after 37, 43 and 49 days at the three treatments. On meat loaves stored at −4 to 7°C, however, no difference between vacuum and N_2 packages were observed.[169] Generally the influence of oxygen exclusion and CO_2 addition is parallel to that on fresh meat. The importance of the Gram negative spoilage flora diminishes and lactobacilli eventually dominate.

11. EFFECT OF SODIUM CHLORIDE

Addition of sodium chloride to meat is the most common way of preservation in order to affect microbial growth and achieve a longer shelf life. Even when comparing different kinds of meat products with different sodium chloride concentrations, important observations can be made. At increasing salt concentrations in the range (salt% in water phase) 4·2–5·8, increasing lag phases and reduced growth rates were observed.[170] The same observations may be made when using different salt concentrations within the same product, i.e. Bologna, sliced salted whole meat product etc. However, great differences of salt tolerance between the spoilage micro-organisms exist. Generally members of *Micrococcaceae* are the most salt tolerant in refrigerated meat products. While these bacteria often constitute a high proportion of the total flora at the beginning of the storage period, they are normally overgrown by other bacteria during storage, especially the lactic acid bacteria, *B. thermosphacta*, etc. However, at increasing salt concentrations, the change in flora occurs at increasingly later stages, or not at all.[40,170-172] This has been observed at 2–10°C, and at moderately high salt concentrations in luncheon meat products. Especially at higher salt levels in bacon, this has also been reported. Comparing bacon with salt concentrations of 5–7% and 8–12% (in the water phase) the *Micrococcaceae* together with *Leuconostoc* dominated at the high level. The same bacteria were found together with streptococci, lactobacilli and pediococci at the lower level.[163] In highly temperature abused packages (30°C compared with 20°C) a change within *Micrococcaceae* from micrococci to coagulase negative staphylococci was observed, regardless of the salt concentration.[175] At salt concentrations of 7–17% in vacuum-packed bacon an approximately linear correlation between log (micrococci/lactics) and salt concentration was observed.[174] Also growth of micrococci may, however, be completely inhibited; thus no development of the flora in vacuum-packed sliced bacon with a salt/water concentration of 10% was observed at 0–1°C during 3 months, a minor development being seen at 1–4°C.[176]

The various lactic acid bacteria are not influenced to the same extent by sodium chloride level. The 'atypical streptobacteria', which are important bacteria in vacuum-packed meat products, are relatively tolerant to salt at the levels used in luncheon meats.[20,177] The relatively low salt tolerance of *Enterobacteriaceae* therefore often makes the lactic acid bacteria constitute a greater proportion of the total flora at increasing salt levels in the lower range. Some lactics are, however, more influenced by increasing salt levels, with longer lag phases and lower maximum numbers,[177,178] and as mentioned the lactics cannot compete with micrococci at the higher salt levels. Experiments with vacuum-packed Bologna showed that temperature was an important factor, i.e. at low temperature growth was restricted even at low salt levels, while only higher salt levels influenced the lactics at higher temperatures.[40]

B. thermosphacta cannot compete with lactics and micrococci at higher salt concentrations in vacuum-packed Bologna (range 3·8–6·4% in water phase), both lag phase and maximum numbers being influenced,[40] although growth was observed at concentrations higher than those normally used in luncheon meats.

The inhibition of *Enterobacteriaceae* was strongly temperature-dependent in studies by Nielsen.[40] In a vacuum-packed whole meat product (range 2·9–4·7% in water phase) development at 10°C was more or less independent of salt concentration, an increasing inhibition being observed at 5°C and 2°C. In a Bologna-type product (range 3·7–6·4% in water phase), the high level limited growth, even at 10°C (4 $\log_{10}$ units increase compared with about 8 $\log_{10}$ units at 3·8%), while no growth occurred at lower temperature.

The influence of salt level on growth of yeast is generally low. Increasing levels in meat products results in yeasts constituting an unchanged or higher proportion of the flora.[172,173] However, lower numbers were observed in vacuum-packed bacon at salt/water ratios above 5.[178]

Atypical *Vibrio*, which may be found in vacuum-packed luncheon meats, were not affected by salt in the usual range for these products.[40] and could constitute a high proportion of the flora at all salt levels, especially at lower temperatures.

The influence of sodium chloride on the normal spoilage flora is enhanced by other factors. No effect of salt in the range 2–5% was observed in non-evacuated bacon; inhibition was only observed when evacuating the product.[179] In bacon stored at 5°C, total counts of $10^{6·75–7·25}$/g were reached after 3 weeks storage at 4% sodium chloride without nitrite addition, at 2·5% salt and 19 ppm nitrite or at 2·0% salt and 48 ppm nitrite,[179] the effect on lactic acid bacteria being much smaller.

At pH values of 6–6·5, growth of *C. botulinum* (A and B) occurs at brine concentrations of up to about 8%.[180] Thus salt as a single factor would not inhibit development in packaged meat products. Meat slurry studies showed that even in the range 2·5–4·5% (brine) increasing salt levels decreases toxin formation.[181] Reducing the temperature would render *C. botulinum* more susceptible to sodium chloride. Studies on Bologna (brine concentration 3·8–6·8%) showed toxic samples (type A and B) at 30°C, but not at lower temperatures, and only at 3·7%.[182] In cooked sliced ham, toxic samples were also found at 4·8% at 30°C and at 3·9% (brine) at 20°C, but not at 15°C.[182] The different results were suggested to be related to a drop in pH during storage of Bologna, but not of ham. *C. botulinum* type E was inhibited at lower salt concentrations (4·4% brine at 20°C). At abusive temperatures, *C. botulinum* could develop at normal salt concentrations in vacuum-packed meat products. In controlling development, one would depend on a low frequency and low levels, as well as low temperatures (type A and B) and factors like nitrite and a competing flora with subsequent drop in pH.

Growth of *S. aureus* is observed at much higher salt concentrations than are normally used in cured meats. However, low temperature enhances the influence of salt. In ham cured in slices and experimentally inoculated with *S. aureus*, toxin was produced at a salt level of 9·2%[136] after 2 weeks at 22°C, while this happened after 16 weeks at 10°C and only up to 4·6% salt.[136] In broth culture, toxin was not produced at 4% sodium chloride and above, at temperatures below 35°C.[183] Another study showed that while 4% salt did not inhibit toxin formation at pH 7·0, decreasing this to 6·5 reduced toxin production, which was inhibited at pH 6.0.[184] At 2%, toxin was also produced at the lower pH levels. In an experiment, where only growth was recorded, sodium chloride in the range 3·4–6·0% (brine concentrations) did not affect growth of *S. aureus* in pure culture on vacuum-packed Bologna at 8–12°C.[185] Thus, while growth may not be inhibited at reasonably low temperatures, the ability to form toxins is more easily restricted.

A temperature-dependent growth inhibition by salt on *B. cereus* was observed in vacuum-packed Bologna.[185] A mixture of five strains showed growth at 3·4% at 10°C; at 12°C they also showed growth at 4·5% (brine), but not at 6·0%. Experiments showed that *B. cereus* grew at 25°C in BHI at this concentration.

Only a slight growth reduction of salmonellae was reported at 2 and 3% salt in broth culture (35°C).[186] Similarly, a brine concentration of 2·7% did not inhibit aerobic growth at 22°C, while 5·7% did.[92] In ground pork with added salt, *Salmonellae* were inhibited at 5% but not at 3·5%.[187] Similar results were obtained in vacuum-packed Bologna, where *S. enteritidis* grew at 12°C at 2·7 and 3·4% (brine), but not at 6·0%.[185] *S. typhimurium* grew at 15°C, at 2·8% but not at 6·0%. In broth culture, the two *Salmonellae* grew at 6·0% at 25°C (*S. enteritidis*) and 15°C (*S. typhimurium*). Reduced growth of *Y. enterocolitica* has been reported in broth cultures at 3°C, with 5% added salt, and death at 7%,[188] while at higher temperature (30°C) growth has been reported at 8 but not 9%, when using high inoculation levels.[189] On the other hand, a strong inhibition was observed in pasteurized pork at 27°C with 3% sodium chloride (decrease of about 3·5 and 2·5 log units for serotype 03a and 08, respectively, compared with control).[190] In Bologna, 6·0% (brine), inhibited at all temperatures from 2 to 12°C, while two strains (type 03) (type 09) were inhibited at 4·5% at 2°C, but not at higher temperatures, and one strain (03) increased about 1 log unit at 4·5% at 2°C. At 3·4% salt, all strains showed profound growth at 2°C.[185]

In packaged meat products, a temperature of 2–5°C and brine concentrations of 3·5–4·5% would render these products free of growth of most pathogens. However, *Y. enterocolitica* could grow under these conditions in pure culture, and one would have to depend on other factors like a competing flora or the absence of these bacteria.

12. EFFECT OF SODIUM NITRITE

A vast amount of research has been done on the effect of sodium nitrite in microbiological media. Experiments with a wide variety of bacteria have shown that the inhibition caused by nitrite is highly influenced by other factors.[90,191,192]

An increasing inhibition is observed at decreasing temperatures, at lower pH and at higher sodium chloride concentrations, i.e. increasingly unfavourable conditions render the micro-organisms more susceptible towards nitrite. The greater effect at decreasing pH is connected with the increasing concentration of undissociated nitrous acid.[192] While a Perigo effect is formed when heating nitrite in media, the effect of sodium nitrite in pasteurised meat products is primarily that of the residual nitrite left after processing. Experiments with bacon have shown an increasing influence of nitrite on total counts in the range 50–200 ppm.[178,193,194] Some studies have, however, shown an inhibitive effect of even small amounts of nitrite (20 ppm),[178] while large levels of nitrite have had no effect in other studies.[195] This may be attributed to different products, manufacturing procedures, the use of ascorbate, differences in composition of initial flora and length of storage of cured product before eventual heating as well as different temperatures during pasteurisation. In unheated (green) bacon produced with nitrite (brine concentrations from 2000–250 ppm) and the same intended salt levels (actual levels 4·9–6·3%), nitrite levels in the bacon were 17–176 ppm.[193] Reducing the nitrite level from 2000 to 1000 ingoing levels resulted in only a slightly faster increase in the aerobic plate count (APC), while reducing the level to 250 ppm, resulted in higher initial counts and counts during storage. Studies on the influence of sodium chloride levels on the nitrite inhibition showed that the inhibition caused by 20 ppm nitrite at 2·4% salt equalled that of 75 ppm nitrite at 1·7% salt (green bacon).[178] Also in heated bacon, nitrite exerts an influence on APC; thus at 4–5°C, 60 ppm nitrite was necessary for a small effect, 120 ppm reduced growth considerably.[198]

The measurable amount of nitrite is greatly altered after mixing with meat and eventual heating. The measurable concentration quickly decreases, in minced meat for instance to 2/3 of the added level,[199] and heating further reduces the nitrite concentration, the depletion being faster at decreasing pH.[200]

On vacuum-packed frankfurters at 4·5–6·3°C, no effect was observed on APC of nitrite addition up to 156 ppm,[201] and only a slightly positive effect was reported on frankfurters stored at 20°C.[202] Neither was any influence observed at 0–150 ppm at 28°C.[203] The studies on both frankfurters and Bologna-type sausage,[74,204] have shown that nitrite addition prolongs the lag phase by up to a week and reduces maximum numbers of the APC often up to several log units at temperatures of 2–10°C. Once growth started, growth rates were only slightly affected. Storage temperature is very important; at 2°C, 100 ppm added nitrite was shown to exert the same inhibition on the APC as 200 ppm, while at 10°C the effect was negligible.[66] Experiments at even higher temperatures (20°C) have shown the same trend.[66,204] Braunschweiger sausages produced with nitrite showed lower counts during storage than sausages without nitrite.[13]

The influence of nitrite on APC is dependent on the composition of the aerobic flora. In broth cultures (at 5–20°C), addition of 100–200 ppm nitrite has only little effect on *B. thermosphacta*.[40,191] In packaged meat products nitrite inhibits both lag phase and maximum numbers.[66,74] Inhibition was dependent on temperature and product type. Thus at 2°C, 200 ppm nitrite (134 ppm residual) totally inhibited *B. thermosphacta* in Bologna sausage and reduced growth in a sliced whole meat product to $2 \cdot 5 \log_{10}$ units compared with an increase of $6-7 \log_{10}$ units without nitrite addition. At 10°C and 20°C, counts of *B. thermosphacta* rose $4-6 \log_{10}$ units with 200 ppm added nitrite. Reducing oxygen permeability of packaging film influenced growth rates in the presence of nitrite only, not in its absence.[66]

Studies with green bacon showed that reducing the nitrite concentration in bacon from 176–144 to 81 ppm did not affect the lactic acid bacteria, while further reducing the level to 17 ppm (residual level in bacon) promoted growth of the lactics.[193] At increasing salt levels less nitrite was necessary to exert the same effect.[178] Added nitrite at 100–200 ppm somewhat inhibited growth of lactics in sliced Bologna,[74] but not in a sliced whole meat product.[66] No interaction of nitrite and gas permeability of film was observed. The lactic acid bacteria are very differently affected by nitrite, and some even reduced the added nitrite

(100 ppm) to a few ppm shortly after packaging.[105,107] An atypical streptobacteria did not seem to be influenced by ordinary nitrite addition.[20]

The influence of nitrite on the micrococci and yeast is limited, whether it is in bacon or pasteurised luncheon meat products.[66,74,178,194] Likewise, the effect of nitrite addition on enterococci is negligible.[178,179]

Enterobacteriaceae are differently influenced by nitrite; thus in inoculation studies 100 ppm added nitrite had no effect on a mixture of *Klebsiellae* at 5 and 8°C in a vacuum-packed meat product.[89,90] In naturally contaminated Bologna, *Enterobacteriaceae* did not develop at 200 ppm added nitrite (2–10°C) nor at 100 ppm at 2°C, and very restricted growth was observed at 100 ppm and 5°C.[74] In a whole meat product, *Enterobacteriaceae* only developed at 10°C and only in the absence of nitrite.[66] The aerobic Gram negative bacteria, *Moraxella/ Moraxella*-like, were not influenced by nitrite in the whole meat product,[66] while they were inhibited in Bologna, especially at low (2°C) temperature.[74] Atypical *Vibrio* spp. were not affected by nitrite,[66] and it has been shown that *Vibrio* may reduce added nitrite.[43]

Many studies have been concerned with the influence of nitrite on pathogenic bacteria, most, however, with *C. botulinum*. Abusive temperatures, which are not normally connected with the storage of luncheon meat products, have often been used. In studies with frankfurter added 50 ppm nitrite and vacuum-packed, toxin formation was delayed at 27°C; increasing the level to 100 ppm further reduced the numbers of toxic samples and 150 ppm completely inhibited it.[203,205] At 20°C, toxicity was observed after 6 days storage of liver sausage (cooked sausage) in the absence of nitrite, while the product remained non-toxic for 14 days with 100 ppm added nitrite, even at 25°C.[206] Also studies with vacuum-packed sliced ham and tongue showed a delay in toxin formation when adding nitrite (type A and B) at 20–30°C, and the effect was accentuated at increasing nitrite levels.[207] In vacuum-packed bacon an influence was observed at the 100 and 200 ppm levels.[208] At 27°C, 0·4% of samples of bacon with 120 ppm nitrite and inoculated with *C. botulinum* type A and B became toxic, compared with 90% of samples without nitrite.[209] Although experiments with high and low pH meat slurries did not clearly elucidate the influence of pH,[181] pH markedly influenced the effect of nitrite in media, and in a study of summer sausage and starter culture, an inhibition of *C. botulinum* was shown at 50 ppm nitrite and dextrose added, with a subsequent drop in pH. However, 150 ppm nitrite did not inhibit development in the absence of dextrose, where pH remained at 5·63.[210] In the experiment with frankfurters,[203] growth of lactic resulted in a pH of 5·0. Also Pivnick and Bird[86] observed an effect of pH on *C. botulinum*, because of growth of lactics. Thus, the added nitrite might inhibit until the concomitant flora produces a pH inhibitive to toxin formation. When examining the influence of nitrite on *C. botulinum*, the inoculation level is important. Thus in one study no toxic samples were found of vacuum-packed bacon with 10^2/g spores at 20°C, while toxic samples were found with 10^4/g spores.[211] In another study, 170 ppm added nitrite inhibited growth of 52/g spores, while 4300/g were not inhibited even at 340 ppm added nitrite.[87]

B. cereus is also influenced by nitrite in meat products. In cooked sausage it was inhibited at 200 ppm nitrite (500 ppm erythorbate), but not at 100 ppm.[212] In vacuum-packed Bologna, only a slightly prolonged lag phase was observed at 75 ppm (10°C), while 150 ppm nitrite resulted in a lag phase of 2 weeks and lower maximum numbers during growth.[213] In a study of another spore forming bacteria, *C. sporogenes*, addition of 156 ppm nitrite completely inhibited growth in vacuum-packed Bologna (15°C), while growth occurred when nitrite was omitted.[138]

Development of *S. aureus* was not influenced by nitrite level in packaged bacon at 15 or 25°C,[208] neither was *S. aureus* affected by nitrite in frankfurters at 7 or 20°C.[202] However, temperature may again be an additional factor in these experiments, depletion being rapid at high temperatures, and low temperatures being inhibitive by themselves. Thus five strains were inhibited at 10°C in vacuum-packed Bologna at 75 ppm nitrite; at 12°C, however, growth occurred at 150 ppm nitrite.[113]

The temperature effect was also observed with *Salmonellae*, where growth occurred in frankfurters with 156 ppm nitrite at 27°C but not at 15°C.[214] In Bologna slow death occurred at 8°C in the presence of nitrite, a slight increase in its absence.[89] In contrast with the results obtained by Rice and Pierson,[214] *S. typhimurium* grew in Bologna at 12°C and 15°C with 75 and 150 ppm nitrite,[213] although with an extended lag phase at the high level (12°C). *S. enteritidis* was more strongly inhibited by 150 ppm added nitrite at 10 and 12°C, numbers increasing only c. 1·5 $\log_{10}$ unit at 10°C. Numbers of *E. coli* rose 1 $\log_{10}$ unit at 42 ppm added nitrite in Bologna, but were inhibited at higher levels.[90] Coliform bacteria were examined in chicken frankfurters produced with 156 ppm nitrite.[215] Growth was restricted in its presence, about $1.5 \log_{10}$ units increase compared with $4.5 \log_{10}$ units in its absence.

The effect of 75 ppm added nitrite on *Y. enterocolitica* (type 03) at 10°C in vacuum-packed Bologna was negligible. At 150 ppm nitrite, growth was somewhat inhibited; maximum numbers were, however, the same.[213] Even at 5°C, addition of 75 ppm nitrite resulted in only slightly lower maximum numbers after one month storage. At 150 ppm, the lag phase was not extended; growth rates were, however, shorter, and maximum counts about $1 \log_{10}$ unit lower. The influence on *Y. enterocolitica* serotype 09 was similar (unpublished results). The inhibition in the vacuum-packed Bologna was considerably smaller than that reported in a meat model system at 27°C (difference of about $3.6 \log_{10}$ units in series with and without 156 ppm nitrite).[190]

The addition of nitrite to meat products may result in changes in the composition of the flora. Generally, increasing nitrite concentrations reduces the proportion of *B. thermosphacta* and *Enterobacteriaceae*, without affecting *Vibrio*. However, at 10–15°C the addition of 100–150 ppm would have little effect on microbial growth. At 2–5°C, the effect would be stronger. The lactic acid bacteria would normally constitute a greater proportion of the flora. However, the effect is influenced by the level of salt. Thus at higher salt concentrations, growth of micrococci is stimulated at the expense of lactics, and at increasing nitrite levels the proportion of lactics decreases and that of micrococci increases.[172] Permeability of packaging film could have an influence on *B. thermosphacta*. The pathogenic flora are not inhibited at higher temperatures, but when temperatures approach minimum values for growth, inhibition is observed.

13. EFFECT OF NITRATE

For many years, nitrate was traditionally added to bacon, and an anti-microbial activity was assumed. A direct effect of nitrate was reported by Ranken,[215a] but others have not reported such an effect. Added nitrate is reduced to nitrite by nitrate reducing bacteria like *Micrococcaceae, Vibrio,* etc.[2] The nitrate therefore acts as a reservoir for nitrite, and a rise in nitrite concentration may be seen, followed by the normal depletion during storage.[116,174] Experiments with back bacon did not reveal a beneficial effect of nitrate.[193] Storage life of vacuum-packed collar bacon was not improved at nitrate concentrations of 196–204 ppm, but there was a slight improvement at the 538–568 ppm level.[217] Also Shaw[194] found a slightly better keeping quality of collar bacon, but not of back bacon, when adding nitrate. In bacon with a high pH (6·0) inclusion of nitrate improved the keeping quality,[217] but the addition of nitrate to bacon with high pH may, however, lead to unacceptably high levels of nitrite.[216] Generally, nitrate addition has resulted in the *Micrococcaceae* constituting a greater proportion of the flora, and the lactics a lesser proportion.[172,174,218] Yeasts are not influenced by nitrate, while the *Enterobacteriaceae* were somewhat inhibited (ibid), which is natural, considering the effect of nitrite upon these bacteria. Experiments with *C. botulinum* in wieners and bacon[205] showed that nitrate was without influence. Nitrate addition alone without simultaneous addition of nitrite could not restrict the development of *C. botulinum*. Thus nitrate addition in packaged meat products is primarily that of a pool of nitrite formation, and the effect therefore reflects that of nitrite. The addition of nitrate would of course also assist in reducing the water activity (a_w) of the product.

14. EFFECT OF ASCORBATE

Sodium ascorbate or sodium isoascorbate are often added to cured meat products, because they reduce nitrosamine formation and balance pigment formation in the products. The addition of ascorbate may, however, interfere with the microbial effect of nitrite. Addition of ascorbate increases the nitrite depletion in meat products.[219] Another factor may be a chelating effect of metal ions essential for repair of nitrite damaged spores,[220] thus enhancing the inhibition caused by nitrite. However, when abusive temperatures are used, after the product has been stored at refrigeration temperatures, addition of ascorbate has a negative influence on the nitrite inhibition.[220] No effect of isoascorbate was observed on recovery of *C. perfringens* spores from pork cured for 2 weeks at 1–4°C followed by 2 weeks at 12·8°C, the pork produced with 0–550 ppm nitrite and 0–200 ppm isoascorbate.[221] In vacuum-packed wieners addition of ascorbate (105 or 665 ppm) did not affect the influence on total counts of 0–150 ppm nitrite (28°C).[203] Neither did 0–2000 ppm ascorbate influence growth of *C. botulinum* in vacuum-packed sliced bacon, but tended to reduce the storage period at ambient temperatures.[208]

Other studies have shown a stimulating influence of the nitrite effect by simultaneous addition of ascorbate. Addition of 500 ppm isoascorbate to naturally contaminated sausage enhanced the effect of 100 or 200 ppm added nitrite.[222] Addition of isoascorbate caused a $0 \cdot 5 - 1 \log_{10}$ unit reduction in total counts, at 21°C at both levels of nitrite. An even greater effect was observed by a simultaneous addition of citric acid, (involving a pH reduction of $0 \cdot 15$ units). Although the nitrite concentration was reduced in the presence of isoascorbate, there was an enhancing influence of 500 or 1000 ppm isoascorbate upon 100 or 200 ppm added nitrite on *B. cereus* and total counts.[212] The single addition of isoascorbate did not influence development of *B. cereus* or total counts.[212]

15. EFFECT OF SORBATE

The current trend for reducing the nitrite addition to meat products has resulted in a search for other additives with an anti-botulinal activity, together with an inhibitive influence on other micro-organisms. Studies with sorbic acid and potassium sorbate in meat products have been encouraging. While the solubility of sorbic acid in water is low that of the salt is quite high. As a fatty acid the general belief is that only the undissociated acid has anti-microbial activity and the proportion of this increases with decreasing pH. Broth culture studies, however, have shown that although the inhibitory activity of the undissociated acid is 10–600 times greater than that of the dissociated acid, this represents more than 50% of the inhibitive activity above pH $6 \cdot 0$,[145] i.e. pH levels often seen in meat products at the beginning of the storage period. Although much of the work has been done on *C. botulinum*, several studies show that there is an effect of sorbate on the total aerobic flora in meat products. Dipping of beef steaks in a 10% potassium sorbate solution prolongs the lag phase of psychrotrophic bacteria, and thus increases the shelf life 2 days.[223] In bacon, whether stored aerobically or anaerobically at 7°C, addition of $0 \cdot 13\%$ or $0 \cdot 26\%$ potassium sorbate caused lower total counts during storage. The effect was also noticeable after 3–4 weeks.[224] Addition of $0 \cdot 26\%$ sorbate also restricted growth on frankfurters at 7–9°C,[204] with maximum counts of about 1 log unit lower than the control, while no effect was seen on bacon stored at abusive temperatures (22°C or 37°C).[209] In cured meat products one would use a combination of sorbate and nitrite because of problems with colour and flavour when using only sorbate. The combination of 40 ppm nitrite and 0·26% sorbate has been tried, and again there is no effect on total counts at abusive temperatures (ibid.). In bacon stored at 0–5°C, the combination resulted in lower counts than with 140 ppm nitrite alone. The effect was enhanced when using a high-barrier film and high vacuum[225] and more pronounced on fatty tissue. On aerobically stored frankfurters (7–9°C), the combination of sorbate and nitrite was no more effective than sorbate (0·26%) or nitrite (40 ppm) alone.[204]

While the study by Ivey et al.[224] showed significantly lower numbers of lactic acid bacteria in bacon with sorbate than in control samples, no consistent influence was reported by Hallerbach and Potter.[204] Neither did Konuma et al.[227] observe any influence on lactics, and the most frequently isolated bacteria by Wagner et al.[225] were lactobacilli and Bacillus.

Potassium sorbate alone inhibited growth of salmonellae in pasteurised sliced turkey at a concentration of 0·12%,[226] at 0·26% in frankfurters (lowest level tried)[214] at 15°C and in sliced Bologna at 0·1% at 10°C, but not at 15°C.[228] C. sporogenes was inhibited equally well by 156 ppm nitrite and 0·26% sorbate in vacuum-packed Bologna at 15°C.[138]

Growth of S. aureus was restricted in the sliced turkey product at 0·12% sorbate and 15°C[226] (1 $\log_{10}$ unit increase compared with 4·5 on control), and in Bologna with 0·1% at 10°C, but not at 15°C.[228] However, another study showed nitrite alone or in combination with sorbate more effective towards S. aureus than sorbate alone.[229] At abusive temperature, growth was only delayed 1 day at 0·1% sorbate.[230] Studies on C. botulinum have shown that a significant reduction of toxic samples of bacon was obtained when adding sorbate.[224,231] In the former study, the addition of 40 ppm nitrite did not enhance the effect of sorbate; however, other studies have shown that a combination of 0·26% sorbate and 80 ppm nitrite is highly efficient, and maintains the same degree of inhibition of C. botulinum as 120 ppm nitrite.[232] A decreased toxin production by sorbate addition has also been reported in studies with poultry frankfurters,[234] chicken emulsions,[233] meat slurries[235] and cooked pork/beef sausage.[230] Often high temperatures are used; however, reducing the temperature from 20°C or 15°C to 10°C, showed that C. botulinum was controlled in blood and liver sausage by 83 ppm nitrite or 0·1 and 0·2% sorbate.[228]

Furthermore, meat slurry studies have shown that increasing the sodium chloride addition or decreasing the temperature enhances the effect of sorbate.[235]

Results of experiments with the normal spoilage flora and pathogens have showed that a good protection against pathogenic bacteria in meat products is obtained with a combination of potassium sorbate and reduced levels of nitrite compared with higher ingoing nitrite levels. The development of the normal flora is influenced at least as much as when nitrite addition is used alone.

16. EFFECT OF MEAT COMPOSITION—pH—PHOSPHATE

16.1. Fresh Meat

In normal meat muscle glycogen is converted to lactic acid after death and the pH decreases to 5·5–5·6. DFD (dark–firm–dry) meat occurs when the meat is deficient in glycogen or the breakdown process post-mortem is abnormal. In both cases this may result in an ultimate pH between pH 6 and as high as 7. Spoilage of DFD meat happens at a faster rate than meat of normal pH. It has been suggested that the difference in pH results in higher growth rates of spoilage organisms on DFD meat. A difference in total bacterial counts has also been observed. Differences of 0·5–1·5 $\log_{10}$ units between normal and DFD meat were reported after vacuum-packed storage at 1–4°C.[236–239] The numbers of lactic acid bacteria were slightly lower on DFD meat (about 0·5 log unit) and also the influence on B. thermosphacta was small in the studies of Hermansen[238] and Erichsen and Molin,[236] while only observed in DFD meat in that of Erichsen et al.[237] Counts of coliforms were more or less equal[238] and that of Enterobacteriaceae 1 $\log_{10}$ unit high on DFD meat.[237] However, numbers of H_2S producing bacteria were strongly affected by the type of meat. These bacteria were either not present on normal meat after storage, or numbers were 5–6 $\log_{10}$ units higher on DFD meat.[236–239] Most of all, the H_2S producing bacteria generally comprise Pseudomonadaceae, especially A. putrefaciens, although H. alvei seemed responsible in the study of Erichsen et al.[237] The spoilage of DFD meat during vacuum-packaged storage may be due to a greening of the meat, with A. putrefaciens being responsible, and unpleasant odours, where S. liquefaciens is important. This could not be prevented even when using very low permeability film,[239] while a good film was effective with meat of pH 5·9–6·1. Growth of A. putrefaciens could be controlled (at −1°C) by treating with citrate buffer.[243] During growth of bacteria on fresh meat glucose is used preferentially and only after this has been exhausted will bacteria attack amino acids and cause putrid odours.[111,112] When glucose is absent on DFD meat, amino acids will be attacked at an earlier stage and spoilage occurs at a faster rate. A gradient of glucose develops in aerobically and anaerobically stored meat, the glucose at the surface being exhausted.[97,112] Supplying glucose at the surface of aerobically incubated meat showed that spoilage by *Pseudomonas* was delayed compared with that of normal meat, while this could not be achieved by a pH reduction.[241] However, other studies on meat (ground meat) showed that samples containing 2–10% glucose had a longer shelf life because the *Pseudomonas* produced lactic acid and pH decreased, which inhibited growth (numbers of lactics were only slightly lower in low glucose meat).[242]

Studies by Gill and Newton[244] have shown that bacteria growing aerobically at low pH do so at much the same growth rates as at higher pH. Thus, aerobically, growth of *Pseudomonas* will not be faster on DFD meat than on normal meat.[111] Similar results on *Pseudomonas* were obtained at pH 5·8–7·0.[245] Other studies were contradictory, here fluorescent *Pseudomonas* grew faster on DFD meat.[246] Furthermore, *A. putrefaciens* would either not develop on low pH meat (5·7–5·8), while being important on DFD meat,[247] or growth rates would be twice as long compared with pH 7.[245] The Gram negative *Moraxella/Acinetobacter* are also inhibited by low pH,[244] and would not develop on refrigerated stored fresh meat even in the presence of oxygen; a temperature increase would diminish the pH effect and so would a pH increase as on DFD meat. Studies by Gill and Newton[244] showed that 20% of *Enterobacteriaceae* were inhibited by low pH. Growth studies showed that *S. liquefaciens* was highly influenced by pH in broth cultures,[40] and while these bacteria may be important on DFD meat,[240,247] they are inhibited on low pH and normal pH meat under anaerobic conditions, or at low oxygen permeabilities.[99] Addition of glucose or citrate has been shown to control *S. liquefaciens* in DFD meat.[247] Studies on other Gram negative bacteria showed similar growth at pH 5·4–5·6 and 6·0–6·3 aerobically on meat, but not on the low pH meat stored anaerobically.[99] With vacuum-packed beef, generation times were higher for *S. liquefaciens* and maximum numbers were lower for *Y. enterocolitica* on low pH than on high pH meat.[247] pH was also important for *B. thermosphacta*, which was inhibited anaerobically on low pH beef and lamb (pH 5·8), while growing at that pH and lower aerobically.[67,134]

Inhibition on low pH meat can be attributed to high concentrations of lactic acid, concentrations which decrease at increasing pH.[104] Broth culture studies have supported these observations; thus growth occurred anaerobically at low pH when adjusting with hydrochloric acid, but not when adjusting with lactic acid. The undissociated lactic acid inhibits anaerobic growth at the low pH.[99,104] Growth of lactic acid bacteria is not enhanced by high pH values in vacuum-packed meat. Actually, higher maximum numbers were found on normal meat (low pH) than DFD meat,[238,247,248] although generation times were equal on both types of meat.

16.2. Effect of pH in Cured Meat

The properties of processed meat may be influenced by pH differences in the fresh meat from which it is derived. Studies of bacterial counts on Wiltshire bacon with pH of 5·7 and 6·5–6·7, respectively, showed[249] a trend towards higher counts on high pH bacon.[217] The storage stability was better on high pH bacon when nitrate was added while this effect was not observed on bacon with pH 6·0.[193] The effect was attributed to increased nitrite concentrations being produced in the bacon at high pH. Other studies showed that the nitrite produced by bacterial reduction of nitrate in vacuum packages may be above the permitted levels during storage at 5°C.[216]

16.3. Glucose in Meat Products

Glucose is not of the same importance for spoilage in vacuum-packed cured cooked meat products as on fresh meat. In studies of vacuum-packed Bologna, it was shown that addition of 1% glucose has negligible influence on growth of B. thermosphacta and S. liquefaciens. Growth of lactobacilli was stimulated at 2°C, but not at 8°C by glucose addition (all experiments made as inoculation studies).[250] In naturally contaminated packages, bacteria were stimulated by glucose in batches with tripolyphosphate added, while no improvement was observed in batches with a low pH phosphate.[251] Addition of glucose resulted in significantly lower organoleptic scores.

16.4. Effect of pH in Cooked Meat Products

Investigations of the effect of pH in cooked meat products are relatively few. In a study of sliced vacuum-packed Bologna at pH 6·07 vs 6·71, no effect was observed on the aerobic mesophile count at 10°C.[14] At 4°C, numbers were highly influenced by pH, with differences of approx. $2 \cdot 5 \log_{10}$ units between batches. No influence was observed on enterococci, and little effect on lactics, as in inoculation studies.[248] The influence on growth of *Enterobacteriaceae* was rather small at 4 and 10°C, (differences of approx. 1 $\log_{10}$ unit after 3 weeks at 10°C). In inoculation studies on vacuum-packed Bologna with pH 6·0 and 6·7, numbers of Gram negative bacteria were higher at the higher pH at 2 and 8°C.[248] Differences of 6·15–6·25 *vs* 6·4–6·5 in vacuum-packed sliced ham resulted in total counts at least 2 $\log_{10}$ units lower at the lower pH.[252]

16.5. Phosphate Addition to Meat Products

Several studies and patents have reported an inhibitive effect upon micro-organisms of phosphate addition to microbial media.[253] In culture media, effects other than that on pH have been observed, i.e. chelation of essential metal ions, etc. Meat products are, however, more complex, and other factors are involved. For instance the added tripolyphosphate quite rapidly hydrolysed to pyro- and orthophosphate by meat enzymes,[254] and orthophosphate has been shown to be without effect at least in media.[255] In vacuum-packed Bologna, considerably reduced growth of *B. thermosphacta* and *S. liquefaciens* at 2°C were observed in samples of low pH (5·9, batch with addition of mixture of sodium pyrophosphate, tripolyphosphate and polyphosphate) compared with pH 6·4 (batches without phosphate or with tripolyphosphate added).[250] Growth rates were somewhat reduced but maximum numbers were little affected. Lactic acid bacteria were not influenced by the pH difference. Addition of low pH phosphate mixture to Bologna somewhat reduced growth of *B. thermosphacta* at 8°C compared with packages where pH was controlled with hydrochloric acid, thus showing an effect of phosphate apart from that of pH. The influence of addition of phosphate in meat products, not resulting in a change in pH was studied with addition of tripolyphosphate or a mixture of tripolyphosphate and pyrophosphate.[250] No effect was observed on *B. thermosphacta,* the aerobic mesophile count or enterococci. Lactics were not influenced in the study by Daelman and van Hoof;[14] growth was somewhat stimulated in series without phosphate addition in that of Nielsen and Zeuthen.[250] Gram negative bacteria were stimulated by phosphate addition in the former study but not in the latter. Higher bacterial numbers were found on matured bacon produced without phosphate than on bacon containing phosphate.[256] In vacuum-packed bacon, however, both initial counts and counts during storage at 5 and 15°C were higher in bacon produced with phosphate.

Studies of the influence of phosphate addition on pathogenic bacteria are few and concern *C. botulinum*. In bacon stored at 27°C, no effect of phosphate addition was observed.[224] Meat slurry studies have shown reduced production of toxin with added polyphosphate.[257] The effect was more pronounced at low pH, and interaction with nitrite was observed. In low pH meat slurries, addition of 0·3% polyphosphate increased pH and toxin formation, while phosphate reduced toxin production in high pH meat slurries.[181,258]

The composition of meat influences microbial growth. In aerobically packaged meat, the DFD condition, with low content of glucose, results in a faster spoilage often reached at lower bacterial numbers because of the development of putrid odours as a result of degradation of low molecular substances. Gram negative bacteria, especially *Pseudomonadaceae* (and here especially *A. putrefaciens*) and *Enterobactericaeae* (*S. liquefaciens* often being the important species) are favoured during storage of DFD meat compared with normal meat. Lactics are more profound on normal meat than on DFD meat. However, pH/lactate has a substantial influence on microbial development in meat. *A. putrefaciens* is highly influenced by pH, and for a number of other bacteria, including *B. thermosphacta,* it has been shown that the lactic acid concentration at the low pH inhibits anaerobic growth. Thus, increasingly anaerobic conditions within a package of low permeability render the bacteria susceptible to lactate/pH. Processed meats are not to the same extent influenced by meat pH as fresh meat, especially when polyphosphate are added. However, in various products an effect on different bacteria has been observed. The influence of phosphate in cured cooked meat products has not been consistent. Effects of phosphate at normal pH are relatively small, while that of phosphate with low pH values may inhibit the spoilage flora. Such an addition could, however, have a negative effect on water holding capacity, while having a positive influence on colour retention.

17. LIGHT–DARK DISPLAY

Generally microbiological experiments have been performed in the dark, while commercial packages for a shorter or longer period are displayed under light conditions. A few studies have been concerned with the effect of light. Packaged fresh beef was not influenced by soft white fluorescent light.[259] Counts were lower on PVC packaged beef at 5°C when displayed under soft white fluorescent light and higher under incandescent light than in the dark.[260] Higher counts were found under light conditions on vacuum-packed veal chucks at −4°C, but not at 0·3 or 7°C,[162] while counts on fresh pork were higher under light.[127] Differences did not exceed $0·5 \log_{10}$ units. Especially at low temperatures, keeping strictly constant temperatures is essential, as displaying under light could lead to a temperature increase and greater microbial development. In studies with meat products, light was not reported to have any influence on microbial growth in meat loaves,[169] or frankfurters.[261] Thus, the influence of light on microbial development in meat and meat products is very small. However, even in the absence of any effect on microbial growth, serious discolorations, off-odours and exudate loss may be the results of storing under light.[162]

18. EFFECT OF FAT AND LEAN PORTIONS OF MEAT

Influence of growth on fatty tissue in packaged fresh meat would not be particularly different from that of the lean portions, because the fat is soaked in moisture from the meat, and thus the low molecular substances would be available for micro-organisms on the fat surfaces. On separate samples of adipose tissue concentrations of glucose, amino acids and lactic acid are, however, lower than on the lean meat of normal pH, and resembling conditions on DFD meat. Spoilage can therefore be influenced by pH and substrate availability in much the same way as on DFD meat. Spoilage was also detected on fatty tissue at a relatively low level, $10^7/cm^2$.[262] This could not be confirmed on aerobically stored pork, where no difference in growth and spoilage between lean and fat was observed.[263] In an atmosphere of CO_2 (89–100%) spoilage was detected after the same storage period, at a time where the aerobic count on the fat was approx. $1·5 \log_{10}$ units lower than on the lean, and thus spoilage at lower microbial levels on fat was observed. *Pseudomonas* dominated on both types of meat, when stored aerobically, with increasing incidence of *A. putrefaciens* and *B. thermosphacta* on fat samples, probably influenced by pH differences (initial pH: lean 5·6, fat 6·8). In CO_2 atmosphere, lactics dominated on both fat and lean. In air-packaged lamb (previously stored under vacuum at 0–1°C) the dominance of *B. thermosphacta* was greater on fat surfaces,[50] (samples obtained by swapping of fat and lean areas of the same package). Incidence of strict aerobic bacteria was greater on lean as was the case in vacuum packages. On naturally contaminated striploins stored at 5°C, (pH of lean 5·5–5·6), growth of *B. thermosphacta, Enterobacteriaceae* and *Pseudomonas* was most profound on fatty tissue, while that of lactics was fastest on lean portions.[264] Furthermore, *Pseudomonas/Moraxella* could not compete with *B. thermosphacta* and *Enterobacteriaceae* on fat, even though growth was faster than on lean. Inoculation experiments were essentially similar, although with equal initial numbers of *Enterobacteriaceae* and *B. thermosphacta* the former outgrew the latter at 5°C, but not at 1°C. *B. thermosphacta* was inhibited at low pH on lean samples (cf. ref. 134). Further experiments showed that *S. typhimurium* and *E. coli* grew at 25°C on fatty tissue and on high pH meat both aerobically and anaerobically.[265] However, aerobically grown cells did not grow on low pH lean meat incubated anaerobically (while anaerobically grown cells grew after an extended lag phase). The author therefore suggested that growth could be more profound during cooling of hot-boned beef on fatty tissue and lean tissue of high pH. Results of growth experiments of other Gram negative bacteria at low temperature are essentially the same.[99] Studies of separately incubated samples of lean and fatty tissue have shown that during aerobic storage the flora is dominated by the aerobic Gram negative flora both on fat and lean. In vacuum packages, growth of *B. thermosphacta* and the Gram negative bacteria is most profound on fatty tissue while lactics dominate on lean portions. However, in mixed samples substrate would also be available on fat surfaces because of drip in the package.

Reports of the influence of fat and lean portions in processed meats are scarce. The lower water content in fatty tissue of bacon would, however, result in a higher brine concentration in these portions, favouring growth of micrococci at the expense of lactobacilli.

19. CONCLUSION

Many factors influence development of micro-organisms in packaged meat. Factors inherent in the product as well as packaging conditions, and for cured, cooked meat products many additional factors, i.e. curing ingredients, affect microbial growth. It has been the intention with this review, not only to point out general aspects of preservation factors, but to focus on the effect on specific groups of bacteria present in meat and meat products, and thereby to describe the possibilities for development of the normal flora and pathogenic bacteria which may be present on the products. It has not been the intention to describe the many different spoilage agents produced during storage, as this has been done for instance in the review by McMeekin,[266] and a detailed study of the microflora associated with bacon was made by Gardner.[2]

REFERENCES

1. GILL, C. O., J. Appl. Bact., 1979, 47, 367–78.
2. GARDNER, G. A., In Meat microbiology, ed. by M. H. Brown, Applied Science Publishers, London, 1982.
3. RIDDLE, W., Technical Cir. No. 404. BFMIRA, 1968.
4. JARMUND, T. and BAARDSEN, B., Rep. No. 34, Norsk Inst. Næringsforskning, 1979.
5. TJABERG, T. A., UNDERDAL, B. and LUNDE, G., J. appl. Bact., 1972, 35, 473–8.
6. CORETTI, K., Fleischwirtschaft, 1978, 58, 1239–1241.
7. WARNECKE, M. O., OCKERMAN, H. W., WEISER, H. H. and CAHILL, V. R., Food Technol. 1966, 20, 118–20.
8. OCKERMAN, H. W., PLIMTON, R. F. and LONG, D. W., Proc. 20th Europ. Meeting Meat Res. Work. 127–129, 1974.
9. NIINIVAARA, F. P., POHJA, M. S. and KREUZER, W., Fleischwirtschaft, 1960, 11, 457–63.
10. WESSELINOFF, W., Berl. München. Tierärzl. Wochenschr. 1957, 70, 516–9.
11. PALUMBO, S. A., HUHTANEN, C. N. and SMITH, J. L., Appl. Microbiol. 1974, 27, 724–32.
12. KEMPTON, A. G. and BOBIER, S. R., Can. J. Microbiol. 1970, 16, 278–97.
13. CHYR, C.-Y., WALKER, H. W. and SEBRANEK, J. G., J. Food Sci. 1980, 45, 1732–5.
14. DAELMAN, W. and VAN HOOF, J., Arch. Lebensmittelhyg. 1975, 26, 213–7.
15. BELL, R. G. and GILL, C. O., J. appl. Bact. 1982, 53, 97–102.
16. BELL, R. G., J. appl. Bact. 1983, 54, 249–55.
17. KITCHELL, A. G. and INRAM, M., 2nd. Int. Cong. Food Sci. Technol. 149–50, 1966.
18. MUKHERJI, S., Dansk Vet. Tidskr. 1976, 59, 832–7.

19. DEMPSTER, J. F. and REID, S. N. *J. Hyg.* 1973, **71**, 815–23.
20. MOL, J. H. H., HIETBRINK, J. E. A., MOLLEN, H. W. M. and VAN TINTEREN, J., *J. Appl. Bact.* 1971, **34**, 377–97.
21. GILBERT, R. J., *J. Hyg.* 1969, **67**, 249–54.
22. SCHMIDT, O. C., *Fleischwirtschaft*, 1961, **13**, 199–200.
23. SIMONSEN, B., *Landbrugsministeriets Slagteri- og Konserveslaboratorium. Rep. No. CH*-5, 1976.
24. SHARPE, M. E., *Food Manufacture*, 1962, **37**, 582–9.
25. BUTTIAUX, R., *Food Manufacture*, 1953, **28**, 135–8.
26. MITRICA, L. and GRANUM, P. E., *Z. Leb. Unter. Forsch.* 1979, **169**, 4–8.
27. INGRAM, M. and HOBBS, B., *J. R. San. Inst.* 1954, **74**, 1151–63.
28. NIVEN, C. P. and EVANS, J. B., *J. Bact.* 1975, **73**, 758–9.
29. BARTELS, H., CORETTI, K. and SCHADECK, I., *Fleischwirtschaft* 1961, **13**, 991–5.
30. BROOKS, R. F. and HENRICKSON, R. L., *Research Bull. 611.* Univ. Missouri. Agric. Exp. Stat. 1956.
31. PARADISO, D. C. and STILES, M. E., *J. Food Prot.* 1978, **41**, 811–5.
32. DUITSCHAEVER, C. L., *J. Food Prot.* 1977, **40**, 382–4.
33. SHAY, B. J., GRAU, F. H., FORD, A. L., EGAN, A. F. and RATCHLIFF, D., *Food Technol. Austr.* 1978, **30**, 48–54.
34. ÅKERSTRAND, K. and NORBERG, P., *Vår föda*, 1980, **32**, 402–8.
35. DELARRAS, C., LABAN, P. and GAYRAL, J. P., *Zbl. Bakt. Hyg., 1. Abt. Orig. B*, 1979, **168**, 377–85.
36. SHAW, B. G. and HARDING, C. D., *J. appl. Bact.*, 1984, **56**, 25–40.
37. KAGERMEIER, A., *Dissert. Ludwig–Maximilian Univ. München*, 1981.
38. REUTER, G., In *Psychrotrophic microorganisms in spoilage and pathogenicity*, ed. by T. A. Roberts, G. Hobbs, J. H. B. Christian and N. Skovgård, Academic Press, London, 1981, pp. 253–8.
39. REUTER, G., *Fleischwirtschaft*, 1970, **50**, 954–62.
40. NIELSEN, H.-J. S., *Ph.D. Thesis*, Technical University, Denmark, 1981.
41. HITCHENER, B. J., EGAN, A. F. and ROGERS, P. J., *J. appl. Bact.* 1982, **52**, 31–7.
42. GARDNER, G. A., *Meat Science*, 1980–81, **5**, 71–81.
43. SHAW, B. G. and LATTY, J. B., *J. appl. Bact.* 1982, **52**, 219–28.
44. ROBERTS, T. A. and SMART, J. L., *J. Food Technol.* 1976, **11**, 229–44.
45. SCHIEMANN, D. A., *J. Food Prot.* 1980, **43**, 360–5.
46. LEISTNER, L., HECHELMANN, H., KASHIWAZAKI, M. and ALBERTZ, R., Fleischwirtschaft, 1975, **55**, 1599–1602.
47. WEISSMAN, M. A. and CARPENTER, J. A., *Appl. Microbiol.* 1969, **17**, 899–902.
48. MYERS, B. R., MARCHALL, R. T., EDMONDSON, J. E. and STRINGER, W. C., *J. Food Prot.* 1982, **45**, 33–7.
49. GARDNER, G. A., CARSON, A. W. and PATTON, J., *J. appl. Bact.* 1967, **30**, 321–33.
50. SHAW, B. G., HARDING, C. D. and TAYLOR, A. A., *J. Food Technol.* 1980, **15**, 397–405.
51. ENFORS, S. O., MOLIN, G., and TERNSTRÖM, A., *J. appl. Bact.* 1979, **47**, 197–208.

52. Stringer, W. C., Bilskie, H. E. and Naumann, H. D., *Food Technol.* 1969, **23**, 97–102.
53. Sutherland, J. P., Patterson, J. T., and Murray, J. G., *J. appl. Bact.*, 1975, **39**, 227–37.
54. Dainty, R. H., Shaw, B. G., Harding, C. D. and Michanie, S., In *Cold tolerant microorganisms in spoilage and the environment*, ed. by A. D. Russel and R. Fuller, Academic Press, London, 1979.
55. Hanna, M. O., Vanderzant, C., Carpenter, Z. L. and Smith, G. C., *J. Food Prot.*, 1977, **40**, 98–100.
56. Christopher, F. M., Vanderzant, C., Carpenter, Z. L. and Smith, G. C., *J. Food Prot.*, 1979, **42**, 323–7.
57. Johnston, R. W., Harris, M. E., Moran, A. B., Krumm, G. W. and Lee, W. H., *J. Food Prot.*, 1982, **45**, 223–8.
58. Seideman, S. C., Vanderzant, C., Hanna, M. O., Carpenter, Z. L. and Smith, G. C., *J. Milk, Food Technol.*, 1975, **38**, 745–53.
59. Beyer, K., *Deutsche Vet. Med. Gesellschaft*, 1979, 158–68.
60. Pierson, M. D., Collins-Thompson, D. L. and Ordal, Z. J., *Food Technol.*, 1970, **24**, 1171–5.
61. Hechelmann, H., Bem, Z., Uchida, K. and Leistner, L., *Fleischwirtschaft*, 1974, **54**, 1515–7.
62. Barlow, J. and Kitchell, A. G., *J. appl. Bact.*, 1966, **29**, 185–8.
63. Collins-Thompson, D. L. and Lopez, G. R., *Can. J. Microbiol.*, 1980, **26**, 1416–21.
64. Alm, F., Erichsen, I. and Molin, N., *Food Technol.*, 1961, **15**, 199–203.
65. Mukherji, S. and Qvist, S., In *Psychrotrophic microorganisms in spoilage and pathogenicity*, ed. by T. A. Roberts, G. Hobbs, J. H. B. Christian and N. Skovgaard, Academic Press, London, 1981.
66. Nielsen, H.-J.S., *J. Food Technol.*, 1983, **18**, 573–85.
67. Egan, A. F. and Grau, F. H., In *Psychrotrophic microorganisms in spoilage and pathogenicity*, ed. by T. A. Roberts, G. Hobbs, J. H. B. Christian and N. Skovgård, Academic Press, London, 1981.
68. Collins-Thompson, D. L. and Lopez, G. R., *Can. Inst. Food Sci. Technol. J.*, 1982, **15**, 307–9.
69. Allen, J. R. and Foster, E. M., *Food Res.* 1960, **25**, 19–25.
70. Luke, K., *Ph.D. Theses*, Freien Univ. Berlin, 1978.
71. Reuter, G., *Arch. Lebensmittelhyg.*, 1970, **21**, 257–64.
72. Steele, J. E. and Stiles, M. E., *J. Food Prot.*, 1981, **44**, 435–9.
73. Gardner, G. and Patton, J., *Proc. 17th Eur. Meet. Meat Res. Work.* 1971, 247–52.
74. Nielsen, H.-J. S., *J. Food Technol.*, 1983, **18**, 371–85.
75. Gardner, G. A., *J. appl. Bact.*, 1979, **47**, iii–iv.
76. Gardner, G. A. and Patton, J., *Proc. 21st Eur. Meet. Meat Res. Work,* 1975, 52–4.
77. Bøgh-Sørensen, L., *Proc. 17th Meet. Meat Res. Work,* 1971, 491–5.
78. Kennedy, J. E., Oblinger, J. L. and West, R. L., *J. Food Sci.* 1980, **45**, 1273–7, 1300.
79. Stiles, M. E., Ng, L.-K. and Paradiso, D. C., *Can. Inst. Food Sci. Technol. J.*, 1979, **12**, 128–30.

80. STILES, M. E. and NG, L.-K., *J. Food Prot.*, 1979, **42**, 464–9.
81. NIELSEN, H.-J. S. and ZEUTHEN, P., *Food Microbiol*, 1984, **1**, 229–43.
82. FARRELL, C. M. and UPTON, M. E., *J. Food Technol.*, 1978, **13**, 15–23.
83. CHRISTIANSEN, L. E. and FOSTER, E. M., *Appl. Microbiol.*, 1965, **13**, 1023–5.
84. GILL, C. O. and NEWTON, K. G., *J. appl. Bact.*, 1980, **49**, 315–23.
85. STILES, M. E. and NG, L.-K., *J. Food Prot.*, 1979, **42**, 624–30.
86. PIVNICK, H. and BIRD, H., *Food Technol.*, 1965, **19**, 1156–63.
87. CHRISTIANSEN, L. N., TOMPKIN, R. B., SHAPARIS, A. B., KUEPER, T. V., JOHNSTON, R. W., KAUTTER, D. A. and KOLARI, O. J., *Appl. Microbiol.*, 1974, **27**, 733–7.
88. MOSSEL, D. A. A., JANSMA, M. and WAART, J. de, In *Psychrotrophic microorganisms in spoilage and pathogenicity*, ed. by T. A. Roberts, G. Hobbs, J. H. B. Christian and N. Skovgård, Academic Press, London, 1981.
89. LEISTNER, L., HECHELMANN, H. and UCHIDA, K., *Fleischwirtschaft*, 1973, **53**, 371–5.
90. LEISTNER, L., HECHELMANN, H., BEM, Z. and ALBERTZ, R., *Fleischwirtschaft*, 1973, **53**, 1751–4.
91. LUITEN, L. S., MARCHELLO, J. A. and DRYDEN, F. D., *J. Food Prot.*, 1982, **45**, 263–7.
92. AKMAN, M. and PARK, R. W. A., *J. Hyg.*, 1974, **72**, 369–77.
93. HANNA, M. O., STEWART, J. C., ZINK, D. L., CARPENTER, Z. L. and VANDERZANT, C., *J. Food Sci.*, 1977, **42**, 1180–4.
94. HANNA, M. O., STEWART, J. C., CARPENTER, Z. L. and VANDERZANT, C., *J. Food Saf.*, 1977, **1**, 29–37.
95. ABDEL-BAR, N. M. and HARRIS, N. D., *J. Food Prot.*, 1984, **47**, 61–4.
96. JENNINGS, M. A. and SHARP, A. E., *Nature*, 1947, **159**, 133–4.
97. NEWTON, K. G. and GILL, C. O., *J. appl. Bact.*, 1978, **44**, 91–5.
98. VELDKAMP, H. and JANNASCH, H. W., *J. Appl. Chem. Biotechnol.*, 1972, **22**, 105–23.
99. GRAU, F. H., *Appl. Environ. Microbiol.*, 1981, **42**, 1043–50.
100. EKLUND, T., *J. appl. Bact.*, 1983, **54**, 383–9.
101. REUTER, G., *Berl. München. Tierärztl. Wochenschr.*, 1962, **75**, 191–3.
102. NIELSEN, H.-J. S. and ZEUTHEN, P., *J. Food Prot.*, 1983, **46**, 1078–83.
103. REUTER, G., *Zbl. Vet. Med. B*, 1972, **19**, 320–34.
104. GRAU, F. H., *Appl. Environ. Microbiol.*, 1980, **40**, 433–6.
105. COLLINS-THOMPSON, D. L. and LOPEZ, G. R., *J. Food. Prot.*, 1981, **44**, 593–5.
106. COLLINS-THOMPSON, D. L., WOOD, D. and JONES, M., *J. Food Prot.*, 1982, **45**, 305–9.
107. DODD, K. L. and COLLINS-THOMPSON, D. L., *J. Food Prot.*, 1984, **47**, 7–10.
108. NIELSEN, H.-J. S. and ZEUTHEN, P., *J. Food Prot.*, 1985, **48**, 28–43.
109. REDDY, S. G., CHEN, M. L. and PATEL, P. J., *J. Food Sci.*, 1975, **40**, 314–8.
110. SMITH, G. C., HALL, L. C. and VANDERZANT, C., *J. Food Prot.*, 1980, **43**, 842–9.

111. GILL, C. O. and NEWTON, K. G., *J. appl. Bact.*, 1977, **43**, 189–95.
112. GILL, C. O., *J. Appl. Bact.*, 1976, **41**, 401–10.
113. MOLIN, G., *Eur. J. Appl. Microbiol. Biotechnol.*, 1983, **18**, 214–7.
114. ATTECK, L. A. G., KIDNEY, A. J. and LOCHE, D. J., *9th Eur. Meet. Meat Res. Work*, 50, 1963.
115. INGRAM, M., *J. appl. Bact.*, 1962, **25**, 259–81.
116. SPENCER, R., *Br. Food Manufac. Ind. Res. Assoc. Res. Rep. No. 136.*, 1967.
117. ENFORS, S.-O. and MOLIN, G., *Meat Sci.*, 1984, **10**, 197–206.
118. KRAFT, A. A. and AYRES, J. C., *Food Technol.*, 1952, **6**, 8–12.
119. BLICKSTAD, E. and MOLIN, G., *J. appl. Bact.*, 1983, **54**, 45–56.
120. AYRES, J. C., *Proc. 18. Ann. Rec. Meat Conf.*, 40–52, 1965.
121. LINDERHOLM, K. G., *Nordisk Hyg. Tidskr.*, 1960, **41**, 17–38.
122. SHANK, J. L. and LUNDQUIST, B. R., *Food Technol.*, 1963, **17**, 1163–6.
123. BARAN, W. L., KRAFT, A. A. and WALKER, H. W., *J. Milk Food Technol.*, 1970, **33**, 77–82.
124. ROTH, L. A. and CLARK, D. S., *Can. J. Microbiol.*, 1975, **21**, 629–32.
125. BOHNSACK, U. and HÖPKE, H.-U., *Fleischwirtschaft*, 1982, **62**, 38–48.
126. WAGNER, M. K., KRAFT, A. A., SEBRANEK, J. G., RUST, R. E. and AMUNDSON, C. M., *J. Food. Prot.*, 1982, **45**, 854–8.
127. IGBENION, J. E., CAHILL, V. R., OCKERMAN, H. W., PARRETT, N. A. and VANSTAVERN, D., *J. Food Sci.*, 1983, **48**, 848–52.
128. NEWTON, K. G. and RIGG, W. J., *J. appl. Bact.*, 1979, **47**, 433–41.
129. JAY, M., KITTAKAA, K. A. and ORDAL, Z. J., *Food Technol.*, 1962, **16**, 95–8.
130. HALLECK, F. E., BALL, C. O. and STIER, E. F., *Food Technol.*, 1958, **12**, 301–6.
131. VANDERZANT, C., HANNA, M. O., EHLERS, J. G., SAVELL, J. W., SMITH, G. C., GRIFFIN, D. B., TERREL R. N., LIND, K. D. and GALLOWAY, D. E., *J. Food Sci.*, 1982, **47**, 1070–9.
132. D'ALLESSANDRIA, A. V. H. and PAGLIARO, A. F., *Fleischwirtschaft*, 1975, **55**, 1582–4.
133. ROTH, L. A. and CLARK, D. S., *Can. J. Microbiol.*, 1972, **18**, 1761–6.
134. CAMPBELL, R. J., EGAN, A. F., GRAU, F. H. and SHAY, B. J., *J. appl. Bact.*, 1979, **47**, 505–9.
135. NIELSEN, H.-J. S., *J. Food Prot.*, 1983, **46**, 693–8.
136. GENIGEORGIS, C., RIEMANN, H. and SADLER, W. W., *J. Food Sci.*, 1969, **34**, 62–8.
137. THATCHER, F. S., ROBINSON, J. and ERDMAN, I., *J. appl. Bact.*, 1962, **25**, 120–4.
138. CHANG, C.-W., SEBRANEK, J. G., WALKER, H. W. and GALLOWAY, D. E., *J. Food Sci.*, 1983, **48**, 861–4.
139. BELL, P. G. and DE LACY, M., *J. appl. Bact.*, 1982, **53**, 407–11.
140. CLARK, D. S. and BURKI, T., *Can. J. Microbiol.*, 1972, **18**, 321–6.
141. WEINFURTNER, F., UHL, A. and OTT, H., *Arch. Microbiol.*, 1960, **36**, 1–22.
142. SHAW, M. K. and NICOL, D. J., *Proc. 15th Meet. Meat Res. Work*, 226–32, 1969.

143. CLARK, D. S. and LENZ, C. P., *Can. Inst. Food Sci. Technol. J.*, 1969, **2**, 72–5.
144. OGILVY, W. S. and AYRES, J. C., *Food Technol.*, 1951, **5**, 300–3.
145. EKLUND, T. and JARMUND, T., *J. appl. Bact.*, 1983, **55**, 119–25.
146. GILL, C. O. and TAN, K. H., *Appl. Environ. Microbiol.*, 1980, **39**, 317–9.
147. SCOTT, W. J., *J. Council Sci. Ind. Res.*, 1938, **11**, 266–77.
148. SUTHERLAND, J. P., PATTERSON, J. T., GIBBS, P. A. and MURRAY, J. G., *J. Food Technol.*, 1977, **12**, 249–55.
149. NEWTON, K. G., HARRISON, J. C. L. and SMITH, K. M., *J. appl. Bact.*, 1977, **43**, 53–9.
150. COYNE, F. P., *J. Soc. Chem. Ind.*, 1932, **51**, 119T–21T.
151. ENFORS, S.-O. and MOLIN, G., *J. appl. Bact.*, 1980, **48**, 409–16.
152. GILL, C. O. and TAN, K. H., *Appl. Environ. Microbiol.*, 1979, **37**, 667–9.
153. ENFORS, S.-O. and MOLIN, G., *Can. J. Microbiol.*, 1981, **27**, 15–9.
154. ENFORS, S.-O., *Köttforskningsinst. Kävlinge, Sweden. Rap. No. 53.* 1979.
155. CHRISTOPHER, F. M., SEIDEMAN, S. C., CARPENTER, Z. L., SMITH, G. C. and VANDERZANT, C., *J. Food Prot.*, 1979, **42**, 240–4.
156. SEIDEMAN, S. C., SMITH, G. C., CARPENTER, Z. L., DUTSON, T. R. and DILL, C. W., *J. Food Sci.*, 1979, **44**, 1036–40.
157. SEIDEMAN, S. C., CARPENTER, Z. L., SMITH, G. C., DILL, C. W. and VANDERZANT, C., *J. Food Prot.*, 1979, **42**, 233–9.
158. SEIDEMAN, S. C., VANDERZANT, C., SMITH, G. C., DILL, C. W. and CARPENTER, Z. L., *J. Food Prot.*, 1980, **43**, 252–8.
158a. HUFFMAN, D. L., DAVIS, K. A., MARPLE, D. N. and McGUIRE, J. A., *J. Food Sci.*, 1975, **40**, 1229–31.
159. POHJA, M. S., ALIVAARA, A. and SORSAVIRTA, O., *Proc. 13th Eur. Meet. Meat Res. Work. B8*, 1967.
160. HUFFMAN, D. L., *J. Food Sci.*, 1974, **39**, 723–5.
161. PARTMAN, W., BOMER, M. T., HAJEK, M., BOHLING, H. and SCHLASZUS, H., *Fleischwirtschaft*, 1977, **7**, 1311–7.
162. LEE, B. H., SIMARD, R. E., LALEYE, L. C. and HOLLEY, R. A., *J. Food Sci.*, 1983, **48**, 1537–42, 1563.
163. SEIDEMAN, S. C., CARPENTER, Z. L., SMITH, G. C., DILL, C. W. and VANDERZANT, C., *J. Food Prot.*, 1979, **42**, 317–22.
164. SPAHL, A., REINECCIUS, G. and TATINI, S., *J. Food Prot.*, 1981, **44**, 670–3.
165. SILLIKER, J. H., WOODRUFF, R. E., LUGG, J. R., WOLFET, S. K. and BROWN, W. D., *Meat Sci.*, 1977, **1**, 195–204.
166. CHRISTOPHER, F. M., CARPENTER, Z. L., DILL, C. W., SMITH, G. C. and VANDERZANT, C., *J. Food Prot.*, 1980, **43**, 259–64.
167. BARAN, W. L., KRAFT, A. A. and WALKER, H. W., *J. Milk Food Technol.*, 1970, **33**, 77–82.
168. OGILVY, W. S. and AYRES, J. C., *Food Res.*, 1953, **18**, 121–30.
169. LEE, N. H., SIMARD, R. E., LALEYE, C. L. and HOLLEY, *J. Food Prot.*, 1984, **47**, 128–33.

170. ALLEN, J. R. and FOSTER, E. M., *Food Res.*, 1960, **25**, 19–25.
171. DEMPSTER, J. F., *J. appl. Bact.*, 1973, **36**, 543–52.
172. HANSEN, N. H. and RIEMANN, H., *Medl. bl.f.D.D.Dryl.for.*, 1961, **44**, 705–24.
173. CAVETT, J. J., *J. appl. Bact.*, 1962, **25**, 282–9.
174. EDDY, B. P. and INGRAM, M., *Proc. 1. Int. Cong. Food Sci. Technol.* 405–11, 1962.
175. TONGE, R. J., BAIRD-PARKER, A. C. and CAVETT, J. J., *J. appl. Bact.*, 1964, **27**, 252–64.
176. HAVAS, F. and TAKÁCS, J., *Fleischwirtschaft*, 1980, **60**, 1276–81.
177. REUTER, G., *Fleischwirtschaft*, 1970, **50**, 954–62.
178. WOOD, J. M. and EVANS, G. G., *Inst. Food Sci. Technol. J.*, 1973, **6**, 111–25.
179. GÓŹDŹ, W. and BORYS, A., *Proc. 21st Eur. Meet. Meat Res. Work,* 159–61, 1975.
180. RIEMANN, H., LEE, W. H. and GENIGEORGIS, C., *J. Milk Food Technol.*, 1972, **35**, 514–23.
181. ROBERTS, T. A., GIBSON, A. M. and ROBINSON, A., *J. Food Technol.*, 1981, **16**, 267–81 and 337–55.
182. PIVNICK, H. and BARNETT, H., *Food Technol.*, 1965, **19**, 1164–7.
183. HOJVAT, S. A. and JACKSON, H., *Can. Inst. Food Technol. J.*, 1969, **2**, 56–9.
184. PEREIRA, J. L., SALZBERG, S. P. and BERGDOLL, M. S., *J. Food Prot.*, 1982, **45**, 1306–9.
185. NIELSEN, H.-J. S. and ZEUTHEN, P., *J. Food Prot.*, 1985, **48**, 150–5.
186. LaROCCO, K. A. and MARTIN, S. E., *J. Food Sci.*, 1981, **46**, 568–70.
187. ALFORD, J. A., and PALUMBO, S. A., *Appl. Microbiol.*, 1969, **17**, 528–32.
188. STERN, N. J., PIERSON, M. D. and KOTULA, A. W., *J. Food Sci.*, 1980, **45**, 64–7.
189. HEIM, D. and HEIM, F., *Ph.D. Thesis,* Humboldt-Univ., Berlin, 1982.
190. RACCACH, M. and HENNINGSEN, E. C., *J. Food Prot.*, 1984, **47**, 354–458.
191. ROBERTS, T. A., BRITTON, C. R. and SHROFF, N. N., In *Food microbiology and technology,* ed. by B. Jarvis, J. H. B. Christian and H. D. Michener, Medicina Viva, Parma. 57–71, 1979.
192. CASTELLANI, A. G. and NIVEN, C. F. JR., *Appl. Microbiol.*, 1955, **3**, 154–9.
193. TAYLOR, A. A. and SHAW, B. G., *J. Food Technol.*, 1975, **10**, 157–67.
194. SHAW, B. G., *Proc. 20th. Eur. Meet. Meat Res. Work.* 114–6, 1974.
195. WIERBICKI, E. and HEILIGMAN, F., *Proc. 26th Eur. Meet. Meat Res. Work.* 198–201, 1980.
196. COLLINS-THOMPSON, D. L. and LOPEZ, G. R., *J. Food Prot.*, 1981, **44**, 593–5.
197. ROBERTS, T. A., JARVIS, B. and RHODES, A. C., *J. Food Technol.*, 1976, **11**, 25–40.
198. HERRING, H. K., *Proc. Meat Ind. Res. Conf. Am. Meat Inst. Found. Chicago,* 47–60, 1973.

199. NORDIN, H. R., *Can. Inst. Food Sci. Technol. J.*, 1969, **2**, 79–85.
200. OLSMAN, W. J., *Voedingsmiddelentechnologie*, 1975, **8**, 7–11.
201. SIMON, S., ELLIS, D. E., MACDONALD, B. D., MILLER, D. G., WALDMAN, R. C. and WESTERBERG, D. O., *J. Food Sci.*, 1973, **38**, 919–23.
202. BAYNES, H. G. and MICHENER, H. D., *Appl. Microbiol.*, 1975, **30**, 844–9.
203. BOWEN, V. G., CERVANY, J. G. and DEIBEL, R. H., *Appl. Microbiol.*, 1974, **27**, 605–6.
204. HALLERBACH, C. M. and POTTER, N. N., *J. Food Prot.*, 1981, **44**, 341–6.
205. HUSTAD, G. O., CERVENY, J. G., TRENK, H., DEIBEL, R. H., KAUTTER, D. A., FAZIO, T., JOHNSTON, R. W. and KOLARI, O. E., *Appl. Microbiol.*, 1973, **26**, 22–6.
206. ALA-HUIKKA, K., NURMI, E., PAJULAHTI, H. and RAEVUORI, M., *Eur. J. Appl. Microbiol.*, 1977, **4**, 145–9.
207. PIVNICK, H., RUBIN, L. J., BARNETT, H. W., NORDIN, H. R., FERGUSON, P. A. and PERRIN, C. H., *Food Technol.*, 1967, **21**, 204–206.
208. CROWTHER, J. S., HOLBROOK, R., BAIRD-PARKER, A. C. and AUSTIN, B. L., *Proc. 2nd Int Symp. Nitrite Meat Prod.* Wageningen *13–20*, 1976.
209. SOFOS, J. N., BUSTA, F. F., BHOTHIPAKSA, K., ALLEN, C. E., ROBACH, M. C. and PAQUETTE, M. W., *J. Food Sci.*, 1980, **45**, 1285–92.
210. CHRISTIANSEN, L. N., TOMPKIN, R. B., SHARAPIS, A. B., JOHNSTON, R. W. and KAUTTER, D. A., *J. Food Sci.*, 1975, **40**, 488–90.
211. COLLINS-THOMPSON, D. L., CHANG, P. C., DAVIDSON, C. M., LARMOND, E. and PIVNICK, H., *J. Food Sci.*, 1974, **39**, 607–9.
212. RAEVUORI, M., *Zbl. Bakt. Hyg. 1. Abt. Orig. B*, 1975, **161**, 280–7.
213. NIELSEN, H.-J. S. and ZEUTHEN, P., *J. Food Technol.*, 1984, **19**, 653–94.
214. RICE, K. M. and PIERSON, M. D., *J. Food Sci.*, 1982, **47**, 1615–7.
215. BAUERMANN, J. F., *Food Technol.*, 1979, **33**, 42–3.
215a. RANKEN, M. D., *J. Sci. Food Agric.*, 1978, **29**, 1089.
216. JOLLEY, P. D., *J. Food Technol.*, 1979, **14**, 81–7.
217. TAYLOR, A. A., SHAW, B. G. and JOLLEY, P. D., *J. Food Technol.*, 1976, **11**, 589–97.
218. SHAW, B. G. and HARDING, C. D., *J. appl. Bact.*, 1978, **45**, 39–47.
219. BEM, Z., HECHELMANN, H. and RAMMING, G., *Bundesanst. Fleischf. Kulmbach. Jahresb.*, 1973, **1**, 49–50.
220. TOMPKIN, R. B., CHRISTIANSEN, L. N. and SHARAPIS, A. B., *J. Food Sci.*, 1979, **44**, 1147–9.
221. SAUTER, E. A., KEMP, J. D., and LANGLOIS, B. E., *J. Food Sci.*, 1977, **42**, 1678–9.
222. NURMI, E., RAEVUORI, M. and HILL, P., *Proc. 21st Eur. Meet. Meat Work.* 209–11, 1975.
223. GREER, G. G., *J. Food Prot.*, 1982, **45**, 82–3.
224. IVEY, F. J., SHAVER, K. J., CHRISTIANSEN, L. N. and TOMPKIN, R. B., *J. Food Prot.*, 1978, **41**, 621–5.
225. WAGNER, M. K., KRAFT, A. A., SEBRANEK, J. G., RUST, R. E. and AMUNDSON, C. M., *J. Food Technol.*, 1982, **45**, 29–32.

226. To, E. C. and Robach, M. C., *J. Food Technol.*, 1980, **15**, 543–7.
227. Konuma, H. and Suzuki, A., *J. Food Hyg. Soc. Japan*, 1974, **15**, 232–51.
228. Leistner, L., Bem, Z. and Hechelmann, H., *Proc. 24th Eur. Meet. Meat Res. Work.* W2.1, 1978.
229. Pierson, M. D., Smoot, L. A. and Stern, N. J., *J. Food Prot.*, 1979, **42**, 302–4.
230. Tompkin, R. B., Christiansen, L. N., Sharapis, A. B. and Bolin, H., *Appl. Microbiol.*, 1974, **28**, 262–4.
231. Huhtanen, C. N., Talley, F. B., Feinberg, J. and Phillips, J. G., *J. Food Sci.*, 1981, **46**, 1796–800.
232. Sofos, J. N., Busta, F. F., Bhothipaksa, K., Allen, C. E., Robach, M. C. and Paquette, M. W., *J. Food Sci.*, 1980, **45**, 1285–92.
233. Sofos, J. N., Busta, F. F. and Allen, C. E., *J. Food Sci.*, 1980, **45**, 7–12.
234. Huhtanen, C. N. and Feinberg, J. J., *J. Food Sci.*, 1980, **45**, 453–7.
235. Roberts, T. A., Gibson, A. M. and Robinson, A., *J. Food Technol.*, 1982, **17**, 307–26.
236. Erichsen, I. and Molin, G., *J. Food Prot.*, 1981, **44**, 866–9.
237. Erichsen, I., Molin, G. and Möller, B.-M., *Proc. 27th Meet. Meat Res. Work.* 783–7, 1981.
238. Hermansen, P., *Proc. 26th Meet. Meat Res. Work.* 300–3, 1980.
239. Tayler, A. A. and Shaw, B. G., *J. Food Technol.*, 1977, **12**, 515–21.
240. Patterson, J. T. and Gibbs, P. A., *J. appl. Bact.*, 1977, **43**, 25–38.
241. Newton, K. G. and Gill, C. O., *Appl. Environ. Microbiol.*, 1978, **36**, 375–6.
242. Shelef, L. A., *J. Food Sci.*, 1977, **42**, 1172–5.
243. Newton, K. G. and Gill, C. O., *J. Food Technol.*, 1980, **15**, 227–34.
244. Gill, C. O. and Newton, K. G., *Appl. Environ. Microbiol.*, 1982, **43**, 284–8.
245. Barnes, E. M. and Impey, C. S., *J. appl. Bact.*, 1968, **31**, 97–107.
246. Rey, C. R., Kraft, A. A., Topel, D. G., Parrish, D. G. Jr. and Hotchkiss, D. K., *J. Food Sci.*, 1976, **41**, 111–6.
247. Gill, C. O. and Newton, K. G., *Appl. Environ. Microbiol.*, 1979, **37**, 362–4.
248. Bem, Z., Hechelmann, H. and Leistner, L., *Fleischwirtschaft*, 1976, **56**, 985–7.
249. Dempster, J. F., *J. Food Technol.*, 1974, **9**, 255–8.
250. Nielsen, H.-J. S. and Zeuthen, P., *J. Food Prot.*, 1983, **46**, 1078–83.
251. Nielsen, H.-J. S. and Zeuthen, P., *Proc. 12th Int. Symp. Microbiol. Ass.* Interactions with food, p. 361.
252. Zeuthen, P., *Symp. Microbiol. Semipreserv. Foods.* Czecholovakia, 1970.
253. Hargreaves, L. L., Wood, J. M. and Jarvis, B., *BFMIRA. Sci. Technol. Surv. No. 76*, 1972.
254. Hamm, R. and Neraal, R., *Z. Leb. Unt. Forsch.*, 1977, **164**, 243–6.
255. Firstenberg-Eden, R., Rowley, D. B. and Shattuck, G. E., *J. Food Sci.*, 1981, **45**, 579–82.

256. KITCHELL, A. G., *Meat Res. Inst. Ann. Rep.,* Bristol, 1970–71.
257. JARVIS, B., PATEL, M. and RHODES, A. C., *J. appl. Bact.,* 1977, **43,** XV.
258. ROBERTS, T. A., GIBSON, A. M. and ROBINSON, A., *J. Food Technol.,* 1981, **16,** 239–69.
259. KRAFT, A. A. and AYRES, J. C., *Food Technol.,* 1954, **8,** 290–5.
260. SATTERLEE, L. D. and HANSMEYER, W., *J. Food Sci.,* 1974, **39,** 305–8.
261. SIMARD, R. E., LEE, B. H., LALEYE, C. L. and HOLLEY, R. A., *J. Food Prot.,* 1983, **46,** 199–205.
262. GILL, C. O. and NEWTON, K. G., *Appl. Environ. Microbiol.,* 1980, **39,** 1076–7.
263. BLICKSTAD, E. and MOLIN, G., *J. Food Prot.,* 1983, **46,** 756–63.
264. GRAU, F. H., *J. Food Sci.,* 1983, **48,** 326–8, 336.
265. GRAU, F. H., *J. Food Sci.,* 1983, **48,** 1700–4.
266. MCMEEKIN, T. A., In *Developments in Food Microbiology,* ed. by R. Davies, Applied Science Publishers, London, 1982.

Chapter 3

CLEANING OF FOOD PROCESSING PLANT

ALAN T. JACKSON

Department of Chemical Engineering, University of Leeds, Leeds, UK

SUMMARY

The fouling of surfaces of food processing plant is a widespread problem, and ranges from surface contamination (which must be removed to prevent cross-contamination of products), the removal of micro-organisms associated with the soil (as a hygiene requirement) to the removal of a severe fouling film (which affects the processing efficiency of equipment). It has been estimated that the cost of fouling heat exchanger surfaces alone throughout the UK processing industries (not only food) is £300–500 million annually. This includes the cost of providing extra heat transfer area to account for the steady loss of efficiency due to the soil deposit, energy costs, maintenance costs and loss of production. This chapter summarises the basic mechanisms affecting food soil formation, methods for evaluating surface cleanliness, the factors which affect soil removal and the practical application of these principles in terms of plant design and the use of modern, automated cleaning-in-place systems.

1. INTRODUCTION

Contamination or soiling of the surfaces of process plant used for food production occurs in many instances, and can range from the simple case of a thin film of material adhering to the surface after draining to the complicated case of a tenacious layer of hard soil of, quite often, denatured material. General plant hygiene considerations dictate that process equipment should be regularly cleaned as part of goodhousekeeping, and the major reasons for cleaning soiled surfaces fall into three main groups.

(a) Where the plant item is used for more than one product, cross-contamination can be eliminated.

(b) Microbiological cross-contamination can be avoided between different batches of the same material.

(c) Periodical removal of soiling films which affect the operating characteristics of the equipment, for example, in heat transfer equipment.

These cleaning problems have some common features, but the ease and efficiency of cleaning and the formulation of the detergent solution used for cleaning are affected by the nature of the soil material. Following food processing operations, surface fouling soils have been found to include residues of one or more of the food ingredients, deposited minerals and impregnated food particles.[1] Food materials are markedly different in their solubility characteristics and their susceptibility to cleaning, particularly if they have been subjected to heating, and Table 1 details some of the major characteristics of a range of food constituents considered by Harper.[2] The cleaning

TABLE 1
FOOD SOIL CHARACTERISTICS

Component on surface	Solubility	Removal characteristics	Change on heating
Sugar	water soluble	easy	caramelisation: more difficult to clean
Fats	water insoluble, alkali soluble	difficult	polymerisation: more difficult to clean
Protein	water insoluble, alkali soluble, slightly acid soluble	very difficult	denaturation: much more difficult to clean
Mineral salts	water solubility variable, most are acid soluble	easy to difficult	generally easy to clean: interaction with other components increases difficulty of cleaning

problems in food processing plants can be arbitrarily classified as follows.

(a) The problem of fouling (and hence cleaning) of surfaces which have been in contact with food materials, but where no physical changes occur in the soil. This situation usually applies where the food material is at ambient temperature and the cleaning operation consists of rinsing the surface to remove lightly adhering material. Frequently in this sort of situation there may be associated bacterial contamination, and the cleaning operation must be designed not only to remove adhering food material, but to reduce the micro-organism count on the surface.

(b) The problem of fouling of surfaces which have been in contact with hot material is a more severe case. Because, for example, the denaturation and insolubilisation of protein in food materials follows a time–temperature relationship, the fact that the hot food material is in contact with the surface for a finite time means that the nature of the components is changing. Hence, pipelines and storage tanks used to handle and hold hot liquids receive a deposited soil as materials in the liquid come out of solution. The nature of this type of soil is different from that due to cold liquids, since the deposited materials bond more firmly to the surface.

(c) The most complicated fouling problem occurs in heat transfer equipment where, because of a deliberate increase in temperature, components of the food material will inevitably be deposited from solution. Unfortunately, because the surface of a heat exchanger is necessarily hotter than the bulk of the liquid, the deposited solids undergo further changes after deposition, usually resulting in a strong soil–surface bond (burning-on).

2. MECHANISMS OF SOIL FORMATION

The mechanisms of soil adhesion or adsorption are complex. Soil can be held to the surface by occlusion in surface irregularities, by electrostatic forces between surface and soil or by the attraction of soil fractions such as minerals and proteins to each other.[3] Sandu and Lund[4] in reviewing the mechanisms of soiling concluded that fouling processes are connected with two types of surface adsorption phenomena—adsorption from complex solutions and adhesion to the surface of small particles. These two phenomena can be classed as chemisorption and/or physical adsorption.

Burton[5] considered soil formation during the heat treatment of whole milk as two separate processes. The elevated temperature produces a condition in the milk in which some of the milk solids are no longer in true solution but are in such a state that they will either agglomerate or adhere to a surface. If a surface is present, these solids will adhere and extend to form a deposit but if no surface is available, they will agglomerate and will no longer be available to adhere to a surface at a later stage.

2.1. Role of Mineral Salts

Burton[5] suggested that a reduction in the solubility of milk mineral salts at elevated temperatures forms the first stage, followed by a slow formation of crystals constituting the soil. Reitzer[6] suggested that where there is only a slight degree of mineral supersaturation spontaneous crystal nucleation is rare and deposit formation takes place after a slow initial nucleus formation and the rate of deposition is governed by the rate of supply of new material.

2.2. Proteins and Lipids

Proteins and lipids have been classed as 'the principal surface-active compounds in nature'[7] and although soil formation is highly correlated with fat content,[8] it is unlikely that the fat itself is closely involved, since the amount of soil does not vary linearly with fat content, and soiling occurs with skim milk as well as whole milk. Burton[8] has suggested that some minor milk constituent associated with the fat and varying with the fat content may be responsible, and phospholipids have received considerable attention.[9–11]

Proteins always seem to be present as a component of food soils, normally in a denatured form, or even degraded into non-protein compounds.[12,13] Prior denaturation of soluble proteins during preheating of milk[5,11] and deliberate pre-heating of tomato juice prior to further thermal treatment[14,15] has been used as a means of reducing fouling.

2.3. Biological Fouling

The adhesion of bacterial and other cells to a solid surface is very complex and occurs throughout all naturally occurring systems as well as the food processing industry. Biological particles cannot adsorb readily to an absolutely clean surface, but surfaces of process plants in contact with fluids containing bacteria, etc. are usually covered with a layer of adsorbed biopolymers. Most cells like bacteria are also surrounded by an envelope of polymeric material, usually carbohydrate, and the adhesion properties of cells are therefore governed by the interaction between the layer of adsorbed polymers on the surface and the cell itself. If, in addition, solids are deposited on the surface due to insolubilisation of mineral salts and proteins, sites become readily available for cell adhesion and subsequent growth. Norde[16] has recently reviewed the physicochemical behaviour of biological materials at solid surfaces showing that the complex behaviour is governed by combinations of Van der Waal's, Coulomb and steric interactions.

3. EVALUATION OF SURFACE CLEANLINESS

Evaluation of cleaning procedures has been limited by the difficulty of determining the degree of soil removal. Various procedures and criteria to evaluate equipment cleanliness and to estimate soil deposit have been developed to meet this demand. Certain advantages are claimed for each technique, but each has its own limitations. Jennings[17] commented that there will probably never be a universally acceptable means of measuring cleaning efficiency.

3.1. Visual Methods
Visual appraisal of the state of cleanliness of equipment surfaces is considered unsatisfactory for quantitative work.[17] In addition, such visual inspections are subjected to reduction in their reliability due to the following factors.[1]

(1) The acuity of vision and perceptiveness in observation vary widely among individuals.
(2) The intensity of lighting—natural or artificial—is usually inadequate making appraisals of cleanliness unreliable.
(3) Films of some product residues and even light encrustations of milk stone are masked when equipment surfaces are wet. Few regulatory inspections are prolonged or postponed so as to provide an opportunity for equipment to dry.
(4) Films of components of some products (proteins, for instance) are not readily detectable visually even when product-contact surfaces are dry.

Ambruster[18] listed nine more reliable visible tests, primarily based on observable physical phenomena, which may be noted either when surfaces are clean or are not clean. These can be grouped as (i) the water break, in which the degree of cleanliness is indicated by the complete sheeting of the rinse water without separating into rivulets, (ii) the droplet test whereby droplets adhere to unclean surfaces, (iii) the salt test utilising salt sprinkled on wet surfaces to render more visible the adhering moisture and (iv) the carbonated water test, whereby gas bubbles adhere to soil films on unclean surfaces.

Several detergent manufacturers have also developed an extremely simple technique called the squeegee-floodlight test to determine the degree of cleanliness. It consists of removal from the surface of any clinging moisture with a squeegee and subsequent drying of the surface by means of a 150 watt, exterior type, sealed beam floodlight. Protein film, otherwise generally undetectable visually, does become visible when subjected to heat. Hair cracks develop and the surface takes on the appearance of weathered aluminium.[1]

3.2. Weighing Methods

The most direct means of measuring soil removal involves accurately weighing soiled test-strips before and after cleaning. This involves adjusting the moisture content of the soil to the same level for each weighing, and can be very time consuming. In addition, a relatively large mass of substrate (metal) can reduce the accuracy of measuring relatively small weight changes in the soil deposited on the test-strips. Hankinson and Carver[19] measured the amount of soil removed by weighing soiled test-pipes before and after circulation cleaning. Cheow and Jackson[20] working on circulation cleaning of a plate heat exchanger weighed the soiled plates before and after cleaning, and found that the reproducibility of this method was well within experimental error.

3.3. Optical Methods

3.3.1. Fluorescent Dyes

Domingo[21] described a fluorochromatic technique for revealing residual soil, particularly on dishes, kitchen utensils and glasses. When the test surface was flooded with water soluble fluorescent dye and observed in darkness under u.v. light, the residual soil consisting of matter possessing fluorescent characteristics was revealed. This method was found to be unsuitable for field application, such as farm milk cooling tanks, transportation and storage tanks, processing vats and pipelines because the dye powder is readily airborne, and adheres to objects in the vicinity causing a red colour problem under damp conditions.[22]

3.3.2. Light-Transmittance

Several workers[23-26] have evaluated cleanliness of glass surfaces by comparing the light transmitted before and after washing. An estimation of the amount of soil removal was based on the Beer–Lambert Law.

3.3.3. Light-Scattering

This technique was developed by A.P.V. Co. Ltd, based on the principle that a light beam is scattered to a degree depending on the amount of soil on the test plate. A constant light source was used to focus onto the test plate at an angle of incidence of 45°. With a clean plate, the light beam is reflected, unscattered, at an angle of 45° and is detected by a light probe.

3.3.4. Turbidimetric Methods

Maxcy and Shahani[27] used both turbidimetric and microbiological procedures to estimate the amount of milk solids picked up by detergent solutions circulated through a pipeline system. Using the Kjeldahl procedure, a degree of correlation was found between the turbidity and nitrogen content of solution. Ruiz[28] later confirmed that the measurement of turbidity of mixtures of milk, detergent and water is largely dependent upon the presence of milk; temperature and detergent (acid and alkali) have only slight effects. The sensitivity of a light spectrophotometer in measuring turbid mixtures was approximately 0·1 mg of milk (about 0·013 mg of milk solids) per ml of the mixture. He concluded that the light spectrophotometer has a very good potential for continuously monitoring the performance of an automated cleaning-in-place (CIP) system.

3.4. Bacteriological Methods

Evaluation of plant and equipment cleanliness is still based on bacteriological analysis with either a contact plate or a swab test. This method provides an index of viable bacteria on the surface but gives no indication of the actual physical cleanliness. A lack of correlation between bacterial count and film deposit was noted by Holland et al.[29]

and they concluded that bacterial counts could not be recommended as a sole measure of cleanliness. A number of workers[27,30-32] have used bacterial count, with or without swab test, for estimating cleanliness. Timperley and Lawson[30] claimed that the swab method which they developed was very reliable, and it was repeatable on the same test after a period of years. Baldock[33] gave a very good review on various procedures for microbiological assessment of bacteria contamination on surfaces of equipment. Unlike the dairy industry, which uses a basis of 10 micro-organisms per square inch, many fruit-based products will contain preservatives and their acidity eliminates the pathogenic organisms. Hence, the usual standard required is physical cleanliness and a microbial count of less than 5 organisms per square inch.[34]

3.5. Radioactive Tracer Techniques

Various methods using a radioisotope tracer to measure the cleanliness of equipment have been suggested. Armbruster and Ridenour[35] and Seiberling and Harper[36] reported that radiological measurements offered greater sensitivity and reproducibility for measuring soil deposits than did other procedures. Cucci[37] used radioactive phosphorus (^{32}P) in studying the removal of milk deposits from rubber, Pyrex and Tygon tubing. However, Seiberling and Harper[36] stated that ^{32}P was unsatisfactory because it reacted irreversibly with stainless steel, while ^{45}Ca appeared to yield more reliable results. In view of the results observed by others and based on laboratory observations, Jennings[38] concluded that probably an adsorption phenomenon between ^{32}P and stainless steel did exist. Jennings et al.[39-42] have used ^{32}P-labelled milk in detergency studies. Because the ^{32}P tracer was generally added as an inorganic salt, it was considered possible that it would leave organic residues unlabelled and might be unsatisfactory for estimating total soil deposits. These later studies reported that there was no difference between in vivo and in vitro ^{32}P-labelled milk, and concluded that addition of the radioisotope gave a suitable index of both organic and inorganic residues. Bourne and Jennings[43] employed ^{14}C-tagged tristearin and ^{14}C-labelled sucrose in their detergency studies. Hays et al.[44] and Masurovsky and Jordan[45] used ^{32}P-labelled bacteria incorporated in a soil as a means of estimating soil deposits. Peters and Calbert[46] followed the rate of soil removal as a function of radioactivity removal for in vivo labelled milk, in vitro labelled milk, and milk containing ^{32}P-labelled bacteria. They concluded that the correlation was best for the latter, the soil being removed faster than radioactivity in the *in vivo* milk and very much faster in the *in vitro* milk. Jennings[47] established a highly significant correlation coefficient of 0·906 between soil removal and radioactivity removal for *in vitro* labelled milk, and attributed the discrepancies in findings to differences in the milk films used by the different investigators.

3.6. Electrical Conductivity

Ruiz *et al.*[48] explored the use of electrical conductivity for monitoring milk residue removal during rinsing. No significant difference was found in the amount of milk residue removed from a stainless steel pipe when the water temperature was either 35°C or 51·7°C and when the flow velocity was varied between 1 and 3 m/s. The sensitivity of electrical conductivity was approximately 0·02 mg milk solids per ml and was limited by ionisation of milk salts when the milk film was dried on the surface. In the study of electrical conductance of milk in the presence of other detergents, Fischer *et al.*[49] concluded that electrical conductance may not be a suitable indicator for milk residues in detergent solutions since more than 75% of the variability in electrical conductance resulted from detergent concentration.

4. MECHANISMS OF SOIL REMOVAL

Food processors, equipment manufacturers and detergent producers have recognised a number of variables (e.g. temperature, concentration, flow rate, time) which influence the cleaning process. Each of these can be varied independently to adjust the cleaning operation for a particular problem or plant operating practice. The conditions generally are selected so as to obtain the best cleaning at least cost. Optimisation for cost-effective cleaning is complex and is a subject demanding detailed in-plant study because of the complex relationships existing between soil loads, choice of detergent, time available, possible maximum flow rate, etc.[3] Considerable work has been directed to establish the relative importance of these factors and to develop means for the evaluation of these variables.

4.1 Effect of Temperature

Conflicting opinions exist regarding the effect of temperature on soil removal. Some imply that high cleaning temperatures encourage the 'burning-on' of residual soils, while others argue that higher temperatures yield better cleaning. Increasing the temperature has the following effects:[2] (1) decreases the strength of bonds between the soil and the surface, (2) decreases viscosity and increases turbulent action, (3) increases the solubility of soluble materials and (4) increases chemical reaction rates. Recommendations of detergent manufacturers and equipment manufacturers range from 40° to 90°C.[50,51] Below 35°C fat remains in a solid state and above 85°C heat-induced interactions occur, binding the protein more tightly to the surface and decreasing cleaning efficiency. For any food soil, the minimum effective temperature will be about 3°C higher than the melting point of the fat. The maximum temperature will depend upon the temperature at which the protein in the system is denatured.[2]

Calbert,[52] using a laboratory circulation unit, reported that pipelines soiled with cold milk could be cleaned by starting with washing solution temperatures of 55–60°C without too much attention to temperature drop during the washing cycle. Smith[53] recommended 40°C for cold-wall milk storage tanks, 77°C for plate heat exchangers used for milk processing and 71°C for milk pipelines. Parker et al.[54] reported that higher temperatures gave more effective cleaning, but Jones[55] stated that temperature has little effect on *detergency* between about 45°C and just below the boiling point of water. These variations have been attributed to the use of ill-defined systems, unsatisfactory evaluation methods and failure to differentiate between the effect of temperature on soil removal and effect of temperature on the detergent form affecting soil removal.[17] The active material of sodium hydroxide solution is probably the hydroxide ion, and the amount of hydroxide ion available is, for all practical purposes, independent of the temperature of the solution.[17] Jennings[41] demonstrated that under the conditions of experimentation, removal of cooked-on milk films by solutions of sodium hydroxide exhibited a Q_{10} of 1·6 over the range 36–82°C, i.e. the rate of soil removal increased by a factor of 1·6 for every 10°C rise in temperature. Jennings[17,38] concluded that the use of kinetic theory, where applicable, is the only approach that will permit measurement and prediction of the effect of temperature (or any other controllable variable) on the cleaning operation with any degree of precision and reliability.

Jackson[56] has shown that the removal of tomato soil from a plate heat exchanger is a complicated process, and that an increase in temperature of 2% sodium hydroxide solution from 20°C to 90°C

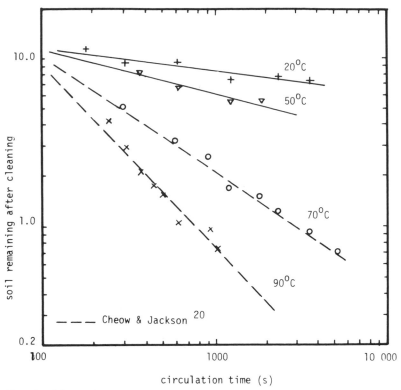

FIG. 1. Effect of temperature on the cleaning of a plate heat exchanger.

decreased the circulation time required and reduced the kinetic order of the cleaning process (Fig. 1). Cheow and Jackson[20] demonstrated that above 70°C, when using only water as the cleaning agent to remove tomato soil from a plate heat exchanger, the rate of cleaning decreases, probably due to protein denaturation.

4.2. Effect of Turbulence

In an industrial situation it is rare that soil deposits can be removed by solubilisation alone, and in almost every case energy must be supplied to effect the final displacement of the soil. For many years the source of this energy was the bristle brush, and in circulation cleaning and spray cleaning, the energy is now applied by friction between the deposited soil and the fluid flowing past it. The shear forces generated at a surface are related to the turbulence of the cleaning solution[38] which can be related to the Reynolds number or fluid velocity.

Early work on turbulence by Jennings *et al.*[39] and Hankinson *et al.*[58] indicated that the Reynolds number was a better criterion than fluid velocity in cleaning pipelines, but work by Timperley and Lawson[30] recently showed that the same degree of cleaning of pipelines of different sizes was not achieved at equivalent Reynolds number but only by using the same mean flow velocity (Fig. 2). The residual micro-organism level on the surface was reduced to a minimum if the flow velocity was about 1·5 m/s, confirming the American 3-A Sanitary Standards recommended flow velocity.[57]

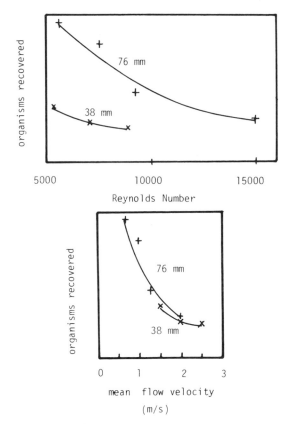

FIG. 2. Effect of Reynolds number and flow velocity on the cleaning of different sized pipes.

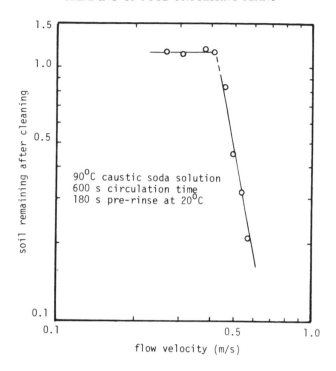

FIG. 3. Effect of flow velocity on the cleaning of a plate heat exchanger.

In the cleaning of a plate heat exchanger fouled by tomato juice, there appears to be a critical flow velocity, below which cleaning is a function of time of contact, but above which a rapid removal of soil takes place for similar contact times.[60] For this particular cleaning situation, an increase in flow velocity above the critical also leads to a lower final residual soil level at constant contact time (Fig. 3).[20,56]

4.3. Effect of Contact Time

Jennings[17] stated it was perhaps misleading to consider 'time' as if it were an independent variable of the cleaning process—time itself make no contribution. Time, therefore, must be considered in relation to some other variables such as temperature, flow rate, etc., i.e. the study of the effect of other variable(s) as influenced by time. Only in a very few restricted cases can soil removal be considered a spontaneous process. Normally most soil-removal mechanisms involve a finite time, and the longer the solution is in contact with the deposited soil under a given set of conditions, the more soil is removed. This reaches a limit because the final traces of soil are probably never removed, and a state where soil redeposition occurs as rapidly as soil removal may be attained. There is both a minimum time for effective cleaning and a practical maximum time for achieving desired results economically.[2]

For a constant flow velocity of detergent solution, the amount of residual soil remaining on the surface of equipment has been shown to be a function of both time of contact and temperature of the detergent solution.[17,20,56,61,62]

4.4 Detergents

4.4.1. Detergent Formulation

Soil becomes attached to surfaces because of an attractive force between the surface and the soil, and its removal is enhanced and accelerated by minimising or neutralising these forces. Harris[64-66] pointed out that one of the primary functions of a detergent is to minimise the soil–substrate attractive forces by adsorption on soil and substrate. 'Detergent' has been defined as 'anything that removes soil'.[17] Bourne and Jennings[67] proposed that a detergent be defined as 'any substance that either alone or in a mixture reduces the work requirement of a cleaning process'. In this definition the word 'work' is restricted to mean mechanical work and is based on the function of a detergent rather than its chemical composition. It specifies no special chemical group, includes soap and other surface-active compounds that possess power of detergency and excludes those that do not. It includes non-surface-active detergents (e.g. sodium hydroxide), synergistic substances (e.g. polyphosphates) and materials that inhibit redeposition (e.g. carboxymethyl cellulose), and solvents and cleaners that degrade soil (e.g. acids), since these reduce the work requirement to zero. It, however, excludes abrasives, since these affect the efficiency with which work is applied, without actually reducing the work requirement in the cleaning operation.

Several workers have attempted to evaluate various types of detergents, the work being reviewed by Jennings,[17] Parry[63] and recently by Tamplin.[3] Schlussler[62] and Harris and Satanek[68] found that surface-active agents were more effective in removing fatty soils than protein deposits. Smith[69] has recommended the use of acid detergents to dissolve heavy metal salts present in protein soil or to remove build-up of milkstone and waterstone, and Kulkani et al.[70] found that an alkaline cleaner was more efficient than an acid one in removing lipids and proteins.

4.4.2. Detergent Concentration

A majority of studies[55,68,71] show that increasing detergent concentration increases detergency to a limit and higher concentrations neither increase nor decrease detergency. This is consistent with the theory that a detergent reduces the forces holding a thin film to a surface but has no further influence above a concentration representing some kind of a saturation effect. Jennings[41] studied the rate of soil removal of radioactively labelled milk films from stainless steel by solutions of $0 \cdot 01 - 0 \cdot 15$ M hydroxide ion. The results showed that the soil removal process was kinetically first order with respect to hydroxide ion as well as soil concentration.

The concentration of detergent required for a cleaning process is generally based on the concentration of active alkalinity or active acidity.[2] For a given detergent there is a minimum concentration for effective cleaning under idealised conditions of water quality and cleaning method. Under industrial conditions, minimal concentrations of compound provide no safety factor for the processing plant. Increasing the detergent concentration where all other cleaning variables are constant, increases cleaning efficiency but at a decreasing rate in exponential decay form. Therefore, there is a maximum concentration for any *practical* improvement in cleaning.

4.5. Effect of Age (Time of Contact between Soil and Surface)

Cleaning recommendations always emphasise that surfaces should be cleaned as soon after soiling as possible. Smith[69] pointed out that unheated sugar and fruits soils can be easily cleaned provided they have not dried. The ageing phenomenon appears to be of general occurrence. In cleaning of cotton fibres by soap solutions, the ageing effect of oil containing soil was attributed to the presence of moisture and polymerisation of unsaturated oil, and Durham[72] comments 'it is well known that when fat is included in the dirt, the longer the soiled fabric is stored before washing the more difficult it is to clean'. Anderson et al.,[73] in studying the removal of tristearin from frosted glass, observed an ageing effect and suggested that this was due to a layer of moisture on the aged discs. However, similar studies by Bourne and Jennings[43,67] presented evidence that tristearin on stainless steel exists in two chemically identical forms that are removed at different rates, and that the ageing effect involves transition of the fast-removal soil species to the more slowly removed form.

4.6. Surface Finish

Kaufmann *et al.*[32,74] studied the cleanability of various surface finishes of stainless steel after soiling with milk inoculated with bacteria. They evaluated the cleaning by swab tests and direct agar contact plate tests and concluded that the surface finishes examined exerted no effect on cleanability as measured by bacteriological tests. In later work, Pflug *et al.*[75] confirmed these findings using radiolabelled soils.

Masurovsky and Jordan[45] demonstrated that the number of ^{32}P-labelled bacteria existing on surfaces after cleaning was affected by the nature of the surface. When three different degrees of surface finish of stainless steel were investigated, retention was less for highly polished surfaces. This result was confirmed by Timperley and Lawson[30] and is shown in Fig. 4.

Stainless steel pipelines can be obtained commercially in highly polished grades, but the surface finish of sheet material for the fabrication of storage tanks and other equipment is not readily obtained in highly polished form. However, recent work on the effects of surface finish indicate that the actual value of surface roughness

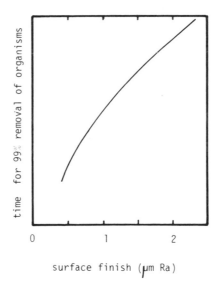

surface finish (μm Ra)

FIG. 4. Time of cleaning *vs* surface finish.[30]

(arithmetic average of surface deviations above and below the mean surface centre line, quoted as μm Ra) is not necessarily important, but rather the shape of the irregularities determines the cleanability.[76] Thus, although electropolishing will show a higher surface roughness than mechanically polished steel, there is no noticeable difference in cleaning. This effect has also been noted in the draining characteristics of storage tanks.[77]

Although it would appear that the surface finish of stainless steel does have some small effect on the cleanability of the surface, other factors such as temperature, flow velocity, etc. play a major part in the cleaning process.

5. PRACTICAL CLEANING APPLICATIONS

In order to successfully and economically clean food processing plant and equipment, the initial design of the plant must take into consideration that not only is a desired process result to be achieved, but the plant must be capable of being satisfactorily cleaned after processing. For example, many different heat exchanger designs and configurations are available for the heating or cooling of liquids, but a number of types cannot be considered hygienic due to crevices and small gaps which would allow material to stagnate and give rise to microbiological contamination or would not allow efficient penetration by cleaning fluids. For food processing applications, these types of heat exchanger are unsuitable, although they are widely used in other parts of the process industries.

5.1. Hygienic Design
The seven basic principles of hygienic design of food plant based on recommendations by the USA 3-A Sanitary Standards Committee, the USA Baking Industry Sanitary Standards Committee and the USA National Canners Association are given below.[78]

(1) All surfaces in contact with food must be inert under the conditions of use, and must not migrate to or be absorbed by the food.
(2) All surfaces in contact with food must be smooth and non-porous so that tiny particles of food, bacteria or insect eggs are not

caught in microscopic surface crevices and become difficult to dislodge, thus becoming a potential source of contamination.

(3) All surfaces in contact with food must be visible for inspection, or the equipment must be readily disassembled for inspection, or it must be demonstrated that routine cleaning procedures eliminate possibility of contamination from bacteria or insects.

(4) All surfaces in contact with food must be readily accessible for manual cleaning, or readily disassembled for manual cleaning, or if CIP techniques are used it must be demonstrated that the results achieved by CIP are equivalent to those obtained by disassembly and manual cleaning.

(5) All interior surfaces in contact with food must be so arranged that the equipment is self-emptying or self-draining.

(6) Equipment must be so designed that the contents are protected from external contamination.

(7) The exterior or non-product contact surfaces should be arranged to prevent harbouring of soils, bacteria or pests in and on the equipment itself as well as in its contact with other equipment, floors, walls or hanging supports.

More detailed recommendations regarding the design of tanks, pumps and pipework have been published by the Joint Technical Committee of the Food Manufacturers Federation and Food Machinery Association.[79]

The design of hygienic food processing plants requires close attention to details, the object being to design a plant which will remain clean during normal operation or be capable of being restored to the desired cleanliness by cleaning. A wide range of equipment is used throughout the food industry, and since it is impossible to discuss all possible cases of hygienic design in detail, some of the more important aspects only will be covered.

5.1.1. Materials of Construction

The materials chosen for the fabrication of food processing plant must meet the requirements of inertness, non-migration or absorption. Since smooth non-porous surfaces lead to hygienic operation during processing as well as relatively easier cleaning (see Section 4.6), certain constructional materials are more suitable than others.

In general the preferred material for food contact surfaces is one of the grades of austenitic stainless steel, in its basic form consisting of

18% nickel, 8% chromium.[80] Where acidic liquids in the presence of salt are to be handled, or where an acidic rinse is required in cleaning, the addition of 3% molybdenum to the steel gives a better resistance to corrosion and possible pitting attack.[81] Both grades of stainless steel can be given a high mechanically polished finish, but electropolishing is quite often satisfactory and cheaper (see Section 4.6).

Aluminium is widely used in the food industry and should be a grade containing minimal quantities of impurities such as zinc and lead. Commercially pure aluminium[82] is essentially 99% aluminium or better, and an anodised finish will help to maintain a smooth hygienic surface over a long period of operation.

The use of plastics and plastic coated metals is satisfactory for low or moderate temperatures, provided that the formulation of the plastic is suitable for food use.[83] Glass lined metals and glass itself can be used, but the thermal and mechanical fragility of glass usually gives rise to problems of potential wholesale contamination of the product, and it is not favoured as a constructional material for food processing plants.

Materials to be avoided include zinc (except galvanised steel for cold water), lead (even in solder—silver solder is safer), cadmium, antimony and plastics containing free phenol, formaldehyde or plasticisers.

5.1.2. Storage Tanks and Vessels

The recommendations regarding the design of storage tanks and vessels detailed in 1967 by the Joint Technical Committee of the FMF and FMA[79] have been supplemented by and incorporated into a number of British Standards, particularly regarding the design of dairy equipment[84,85] and these general principles should be incorporated into the design and construction of storage tanks and processing vessels for all food plants.

The basic principles involved may be summarised as follows and are shown in Fig. 5.

(a) All outlets should be finished flush inside the vessel.
(b) The equipment should be self-emptying or self-draining with no internal pockets.
(c) Lids or covers should not cause drainage into the tank when opened.
(d) Exterior surfaces should be readily cleanable and should prevent accumulation of dirt, bacteria and pests.

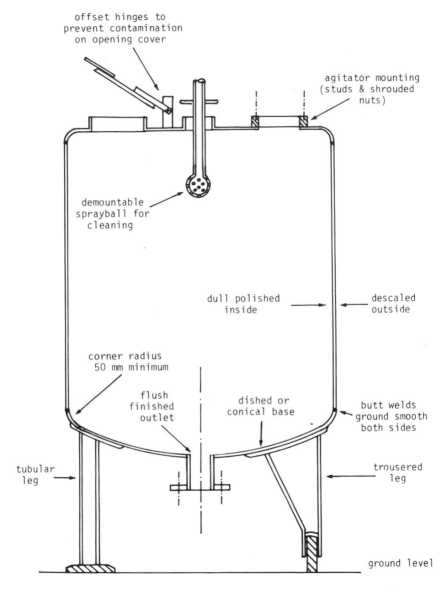

FIG. 5. Hygienic design principles applied to storage tanks.[79]

(e) Shaft seals and bearings should be easily removable for inspection.
(f) All internal corners should be radiused with a minimum radius of 50 mm.
(g) All welds should be butt-jointed and ground smooth.

5.1.3. Pipelines and Fittings
A number of different pipe fittings have been specially designed and developed for use in the food industry, the early designs being specifically developed for the dairy industry. The major consideration to be taken into account in choosing a fitting for joining two lengths of pipe or for fitting pipelines to vessels and other process plants depends upon whether or not the joint is required to be dismantled to allow manual cleaning or whether the major cleaning operation is CIP.

The recessed ring type of joint[86] (Fig. 6) was designed for ease of dismantling, particularly by unskilled operators, and is used where a pipeline or piece of equipment is dismantled for cleaning and sterilising after each processing run. The tightened joint is unsuitable for CIP applications since a crevice is formed at the joint, making it difficult to obtain complete penetration by the cleaning solution. The IDF (International Dairy Federation) joint[88] (Fig. 7) has now been adopted as the standard crevice-free joint for CIP applications, since the seal forms a flush surface with the interior of the pipeline. It is recommended that the fitting be welded to the pipe rather than expanded, the expanded version tending to leave the joint ring protruding into the pipe interior due to the slight increase in diameter of the pipe from the expansion operation (Fig. 8). The use of IDF couplings allows relatively easy dismantling for occasional inspection and manual cleaning.

In large-scale modern plants, lengths of pipework are joined by orbital welding on-site, and equipment has been developed to carry out this operation to give welds of such a high standard that no

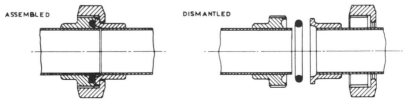

Fig. 6. The recessed ring type joint (RJT) (BSS 1864).

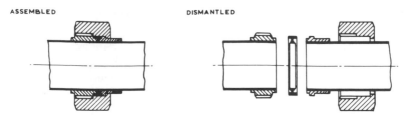

FIG. 7. The International Dairy Federation (IDF) joint (BSS 4825).

grinding or polishing of the weld is necessary. These joints are completely hygienic and are suitable for CIP techniques since no crevices or intrusions appear in the pipe bore in a correctly welded joint.

The joint rings used in the RJT and IDF systems are available in a wide range of food grade rubbers and plastics, but in the rare cases where the use of an elastomeric seal is prohibited (either by temperature or corrosive conditions) it is possible to use the metal-to-metal cone seal.[87] This is a USA 3-A standards developed joint, now incorporated into British Standards, and it is also possible to obtain special versions incorporating an elastomeric seal if required.

In all cases of pipeline erection and manufacture, the correct pipeline for food use must be specified (BS 4825[88]). Stainless steel pipe manufactured to this standard is now usually unpolished, improvements in manufacture giving a generally acceptable finish. It is important in order to ensure satisfactory cleaning, however, to specify that the surface finish conforms to the British and ISO standards for food use, i.e. the surface finish must be less than $1\cdot0\,\mu$m Ra.

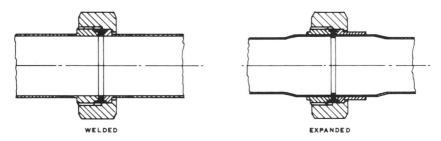

WELDED   EXPANDED

FIG. 8. Welded and expanded IDF joints.

5.1.4. Miscellaneous

Valves, pumps and other ancillary equipment used for food processing applications should be chosen, or designed, to follow the general rules of hygienic operation. Passage shapes should be smooth and have a good surface finish, the equipment should be self-draining and contain no dead pockets or crevices. Clearances between adjacent parts, if a necessary part of the design, should be large enough to ensure adequate penetration by CIP or other cleaning techniques. If CIP cannot be used, then the equipment must be capable of being easily dismantled for cleaning. The use of screw threads in contact with the food material should be avoided, or if this is not possible, only coarse, shallow threads should be used to permit easy cleaning. Any leakages which might occur should not occur into closed spaces (pump motor casings, valve bodies) and should be readily detectable. Joints and gaskets, where used, should form a smooth surface within the equipment and should be positively retained.

External surfaces should also be smooth and be capable of being easily cleaned by simple hosing down. Motors should be of smooth case construction (not ribbed) and all electrical equipment must be hoseproof. Unless equipment is sealed to the floor, it must have a ground clearance of at least 100 mm to ensure adequate cleaning access. Legs for supporting equipment should preferably be tubular with rounded ends and be sealed. Any flat surface should be given a slight outward fall to ensure complete drainage.

5.2. CIP

CIP offers numerous advantages over strip-down cleaning. With a properly designed system, CIP not only gives the obvious advantage of saving time and labour in dismantling equipment, but because the CIP system is usually automatically controlled, a saving in the use of materials (water, detergent, acid, etc.) is also possible compared to a manual cleaning operation which is difficult to control precisely.

CIP systems were originally designed on the basis of installing a centralised plant complete with all necessary pumps and heating equipment, capable of dealing with numerous items of processing plant. An advantage of this centralised system is that the preparation of the detergent, alkali and acid solutions is divorced from the processing plant area. The major drawback of this type of system is in the complicated, long lengths of pipework required to feed each part of the processing plant, and the dilution of the detergent and other

ALAN T. JACKSON

TABLE 2
VOLUMETRIC CAPACITY OF PIPES

Diameter	Volume ($litres/100\,m$)
38 mm ($1\frac{1}{2}$ in)	113
51 mm (2 in)	204
63·5 mm ($2\frac{1}{2}$ in)	316
76 mm (3 in)	454
101·6 mm (4 in)	810

treatment chemicals which occurs due to the large volumes of pre- and post-rinse water contained in the pipelines. Since the fluid contained in the pipes can only be displaced by the solution following in the cleaning process, this can lead to considerable quantities of water returning to the central feed tank. Table 2 shows the volume of liquid contained in a 100 m run of pipe for different pipe diameters.

The current trend is to install a partly centralised system employing self-contained satellite cleaning units. The making-up of detergent, alkali, acid solutions etc. is still carried out centrally, but the plant used for the cleaning cycle employs a small packaged unit. The unit is mounted close to the item or items to be cleaned and consists of a small capacity (1250 litres) tank, pump and heating equipment. The unit is connected to a detergent main running from the centralised preparation area (and may be connected to a supply of other treatment chemicals if required). Water, steam and electricity are connected locally, and if possible a supply of compressed air is used for purging the cleaning lines. A number of these satellite units can be installed throughout the factory, and since each unit contains its own control panel, the cleaning cycle can be tailored to suit each cleaning application.

The general operating cycle for a satellite unit is as follows. (1) The tank is filled with mains water which is then circulated as a pre-rinse through the plant being cleaned. The pre-rinse may be used cold or heated depending on requirements. After the pre-rinse, the water is drained from the plant back to the satellite tank—assisted by using compressed air—and then normally discharged to drain. (2) The satellite tank is filled with detergent or other treatment solution, and this is then circulated through the item being cleaned—again either cold or heated. (3) The chemical solution after circulation is drained back to the satellite tank, and may then be returned to the centralised preparation area, or discharged to drain. (4) The post-rinse operation is carried out in the same manner described in (1) above, this water either being discharged to drain, or saved and used for the pre-rinse of the next cycle.

The only long runs of pipe required in the satellite system are ring mains carrying treatment solutions; the cleaning lines to the processing plant are very short. Although the capital costs of a satellite system are slightly higher than for a completely centralised system, the savings in materials, heating and pumping costs can be as high as 60% of those of a centralised system, and the extra capital cost can be recovered in 1–2 years. The system also lends itself to relatively easy expansion simply by buying an extra satellite unit and connecting it to the services on-site.

CIP of processing plant surfaces can be split into two basic types of operation. (1) Cleaning of large surfaces (tanks, vessels, etc.) where the flow channel is exceptionally large. (2) Cleaning of pipelines where the cross-sectional area is relatively small.

In the cleaning of tanks, etc., where a large volume of solution would be required in order to fill completely the cross-section, the technique used is to flood the surface with cleaning solution using directed jets. A number of devices have been developed to achieve a total wetting of the whole surface, ranging from the simple sprayball (Fig. 9) to the rotating jet system designed to not only wet the surface

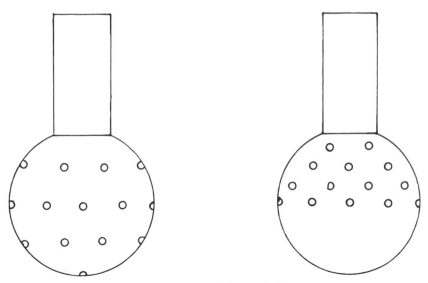

FIG. 9. Typical sprayballs.

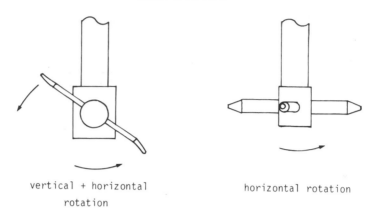

vertical + horizontal   horizontal rotation
        rotation

FIG. 10. Typical rotating spray turbines.

but to give some impingement force between the liquid and surface
(Fig. 10).

For pipelines and other equipment with small cross-sectional flow
areas, the problem is eased, and cleaning is carried out by simply
circulating liquid through the pipe or piece of equipment at an
adequate velocity (at least 1·5 m/s). Provided that the system has been
designed for CIP, this method has proved satisfactory.

5.2.1. Single-Use Systems (Fig. 11)
This is the simplest CIP system possible. Detergent and other
treatment solutions are used once only and then discharged to drain.
Similarly, both the pre- and post-rinse water is used once only. This
type of system is suitable for very heavy soil loadings where contamination of the cleaning fluids is such that recovery is not economic.

5.2.2. Re-Use Systems (Fig. 12)
This type of system is designed to recover and re-use the cleaning
solutions by returning them to the centralised making-up area after
use, where they are filtered, made up to strength and stored ready for
the next cleaning cycle. The re-use of detergents has been the subject
of extensive investigation[89–91] and has been shown to be satisfactory
for moderate soil loads.

This type of system requires a central storage tank for each of the

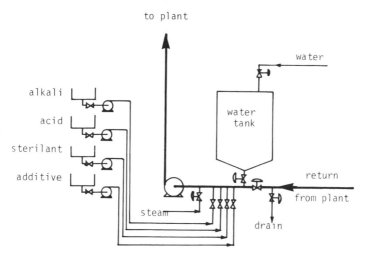

FIG. 11. Single-use CIP system.

cleaning fluids (detergent, alkali, acid, hot water, etc.) and shows considerable chemicals cost savings over the single-use system. The capital costs are usually high due to the need to store all the treatment chemicals for re-use, and for cleaning different process plants the piping system can become very complex.

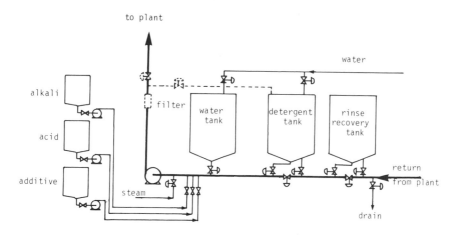

FIG. 12. Re-use CIP system.

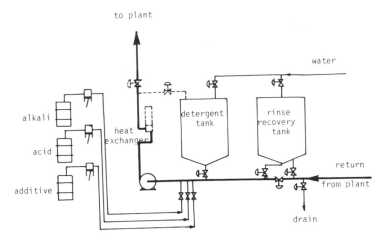

FIG. 13. Multi-use CIP system.

5.2.3. Multi-Use Systems (Fig. 13)

This type of system is designed to reduce the number of storage tanks required for the solutions and basically consists of two storage tanks (detergent and rinse recovery). Detergent strength is maintained by injecting concentrated detergent solution into the circuit using a metering pump, the detergent tank being used as a buffer store. The rinse recovery tank is used initially for the pre-rinse (this water being discharged to drain after use), the post-rinse water being held in store for the next cleaning cycle pre-rinse. Any other chemical treatment is carried out by injection into the circulation pipework (since both major storage tanks may be full).

A detailed operating guide and full description of these three systems has been given by Timperley.[92] It is of course important that the cleaning circuit and processing circuit can be positively isolated from each other, and use is made of blanking plates and isolating valves to ensure no intermixing of cleaning fluids with products.

REFERENCES

1. ABELE, C. A., *J. Milk Food Technol.*, 1965, **28,** 257.
2. HARPER, W. J., In *Food Sanitation* (ed. by R. E. Guthrie) AVI, Westport, 1974, p. 130.
3. TAMPLIN, T. C., In *Hygienic Design and Operation of Food Plant* (ed. by R. Jowitt), Ellis Horwood, Chichester, 1980, p. 183.

4. SANDU, C. and LUND, D. B., *International Conference on the Fouling of Heat Transfer Equipment*, Rensselar Polytechnic Inst., New York, 1979.
5. BURTON, H., *J. Dairy Res.*, 1968, **35**, 317.
6. REITZER, B. J., *Ind. Eng. Chem. Proc. Res. Dev.*, 1964, **3**, 345.
7. MACRITCHIE, F., In *Advances in Protein Chemistry*. 9 (ed. by C. B. Anfinsen, J. T. Edsall and F. M. Richards) Academic Press, London, 1978.
8. BURTON, H., *J. Dairy Res.*, 1967, **34**, 137.
9. ITO, R. and NAKANISHI, T., *Jap. J. Dairy Sci.*, 1966, **15**, A69.
10. HOLDEN, T. F., ACETO, N. C., DELLAMONICA, E. S. and CALHOUN, M. J., *J. Dairy Sci.*, 1966, **49**, 346.
11. ITO, R. and NAKANISHI, T., *Jap. J. Dairy Sci.*, 1967, **16**, A62.
12. ITO, R. and NAKANISHI, T., *Jap. J. Dairy Sci.*, 1964, **13**, A202.
13. NAKANISHI, T. and ITO, R., *17th Int. Dairy Congr.*, Munich, 1966, **B**, 613.
14. ADAM, H. W., NELSON, A. I. and LEGAULT, R. R., *Food Technol.*, 1955, **9**, 354.
15. MORGAN, A. I. JR. and WASSERMAN, T., *Food Technol.*, 1959, **13**, 691.
16. NORDE, W., In *Fundamentals and Applications of Surface Phenomena Associated with Fouling and Cleaning in Food Processing* (ed. by B. Hallstrom, D. B. Lund and Ch. Tragardh), Lund University, 1981, p. 148.
17. JENNINGS, W. G., In *Advances in Food Research* (ed. by C. C. Chichester, E. M. Mrak and G. F. Stewart) Academic Press, London, 1964, p. 325.
18. ARMBRUSTER, E. H., *The Sanitarian*, 1962, July/August.
19. HANKINSON, D. J. and CARVER, C. E., *J. Dairy Sci.*, 1968, **51**, 1761.
20. CHEOW, C. S. and JACKSON, A. T., *J. Food Technol.*, 1982, **17**, 417 and 430.
21. DOMINGO, E., *27th Ann. Meeting Chem. Specialities Mfg Assoc. Inc.*, New York, 1950.
22. BECK, W. J., *Ass. Food Drug Officials U.S.*, 1962, April.
23. JENSEN, J. M., *J. Dairy Sci.*, 1946, **29**, 453.
24. GILCREAS, F. W. and O'BRIEN, J. E., *Am. J. Public Health*, 1941, **31**, 143.
25. MANN, E. H. and RUCHHOF, C. C., *U.S. Public Health Rep.*, 1946, **61**, 877.
26. LEEHERTS, L. O., PIETZ, J. F. and ELLIOTT, J., *J. Am. Oil Chem. Soc.*, 1956, **33**, 119.
27. MAXCY, R. B. and SHAHANI, K. M., *J. Milk Food Technol.*, 1961, **24**, 122.
28. RUIZ, E. L., *Ph.D. Thesis*, University of Missouri, Columbia, 1972.
29. HOLLAND, R. F., SHAUL, J. D., THEOKAS, D. A. and WINDLAN, H. M., *Food Eng.*, 1953, **25(5)**, 75.
30. TIMPERLEY, D. A. and LAWSON, G. B., In *Hygienic Design and Operation of Food Plant* (ed. by R. Jowitt) Ellis Horwood, Chichester, 1980, p. 79.
31. PARKER, M. E. and LITCHFIELD, J. H., *Food Plant Sanitation*, Reinhold, New York, 1962.
32. KAUFMANN, O. W., HEDRICK, T. I., PFLUG, I. J. and PHEIL, C. G., *J. Milk Food Technol.*, 1960, **23**, 377.

33. BALDOCK, J. D., *J. Milk Food Technol.*, 1974, **37**, 361.
34. ROBBINS, R. H., Beecham Products Ltd, Coleford—private communication.
35. ARMBRUSTER, E. H. and RIDENOUR, G. M., *Soap Sanitation Chem.*, 1952, **28(6)**, 83.
36. SEIBERLING, D. A. and HARPER, W. J., *J. Dairy Sci.*, 1956, **39**, 919.
37. CUCCI, M. W., *J. Milk Food Technol.*, 1954, **17**, 322.
38. JENNINGS, W. G., *Food Technol.*, 1963, **17(7)**, 53.
39. JENNINGS, W. G., MCKILLOP, A. A. and LUICK, J. R., *J. Dairy Sci.*, 1957, **40**, 1471.
40. JENNINGS, W. G., *J. Dairy Sci.*, 1959, **42**, 476.
41. JENNINGS, W. G., *J. Dairy Sci.*, 1959, **42**, 1763.
42. JENNINGS, W. G., *Food Technol.*, 1960, **14**, 591.
43. BOURNE, M. C. and JENNINGS, W. G., *Food Technol.*, 1961, **15**, 495.
44. HAYS, G. L., BURROUGHS, J. D. and JOHNS, D. H., *J. Milk Food Technol.*, 1958, **21**, 68.
45. MASUROVSKI, E. B. and JORDAN, W. K., *J. Dairy Sci.*, 1960, **43**, 1545.
46. PETERS, J. J. and CALBERT, H. E., *J. Dairy Sci.*, 1960, **43**, 857.
47. JENNINGS, W. G., *J. Dairy Sci.*, 1961, **44**, 258.
48. RUIZ, E. L., BROOKER, D. B., ANDERSON, M. E. and MARSHALL, R. T., *J. Milk Food Technol.*, 1972, **35**, 257.
49. FISCHER, J. R., BROOKER, D. B., ANDERSON, M. E., MARSHALL, R. T. and RUIZ, E. L., *J. Dairy Sci.*, 1973, **56**, 1405.
50. Reddish Chemical Co. Ltd., Cheadle, *Detergents and Sanitisers for the Food Industry*, Leaflet No. 162, 1978.
51. DAOUD, I. S., APV Co. Ltd., Crawley—private communication.
52. CALBERT, H. E., *J. Milk Food Technol.*, 1958, **21**, 12.
53. SMITH, G. A., *Am. Milk Rev.*, 1957, **19(10)**, 36 and 48.
54. PARKER, R. B., ELLIKER, P. R., NELSON, G. T., RICHARDSON, G. A. and WILSTER, G. H., *Food Eng.*, 1957, **25(1)**, 82.
55. JONES, T. G., In *Surface Activity and Detergency* (ed. by K. Durham), Macmillan, London, 1961.
56. JACKSON, A. T., *1st UK National Heat Transfer Conf.*, *I. Chem. E. Symp. Ser. No. 86*, p. 465, Leeds, 1984.
57. Int. Ass. Milk Fd Env. Sanitarians, *J. Milk Food Technol.*, 1966, **29**, 95.
58. HANKINSON, D. J., CARVER, C. E., CHONG, K. P. and GORDON, K. P., *J. Milk Food Technol.*, 1965, **28**, 377.
59. LUND, D. B. and SANDU, C., In *Fundamentals and Applications of Surface Phenomena Associated with Fouling and Cleaning in Food Processing* (ed. by B. Hallstrom, D. B. Lund and Ch. Tragardh) Lund University, 1981, p. 27.
60. JACKSON, A. T. and LOW, W. M., *J. Food Technol.*, 1982, **17**, 745.
61. SCHLUSSLER, H., *Milchwissenschaft*, 1970, **25(3)**, 133.
62. SCHLUSSLER, H., *Brauwissenschaft*, 1976, **29(9)**, 263.
63. PARRY, D., *J. Inst. Brewing*, 1970, **76**, 443.
64. HARRIS, J. C., *Soap Chem. Spec.*, 1958, **34(12)**, 59.
65. HARRIS, J. C., *Text. Res. J.*, 1958, **28**, 912.
66. HARRIS, J. C., *Soap Chem. Spec.*, 1959, **35(1)**, 49.

67. BOURNE, M. C. and JENNINGS, W. G., *J. Am. Oil Chem. Soc.*, 1963, **40**, 212.
68. HARRIS, J. C. and SATANEK, J., *J. Am. Oil Chem. Soc.*, 1961, **38**, 169 and 244.
69. SMITH, H. W. N., *Candy Ind. Confectioners J.*, 1969, **132(4)**, 22 and 25.
70. KULKANI, S. M., ARNOLD, R. G. and MAXCY, R. B., *J. Dairy Sci.*, 1975, **58**, 1095.
71. KLING, W. and LANGE, H., *Kolloid Z.*, 1955, **142**, 1.
72. DURHAM, K., In *Surface Activity and Detergency* (ed. by K. Durham), Macmillan, London, 1961.
73. ANDERSON, R. M., SATANEK, J. and HARRIS, J. C., *J. Am. Oil Chem. Soc.*, 1959, **36**, 286.
74. KAUFMANN, O. W., HEDRICK, T. I., PFLUG, I. J., PHEIL, C. G. and KEPPELER, R. A., *J. Dairy Sci.*, 1960, **43**, 28.
75. PFLUG, I. J., HEDRICK, T. I., KAUFMANN, O. W., KEPPELER, R. A. and PHEIL, C. G., *J. Milk Food Technol.*, 1961, **24**, 390.
76. MILLEDGE, J. J. and JOWITT, R., *I.F.S.T. Proc.*, 1980, **13(1)**, 57.
77. MILLEDGE, J. J., In *Fundamentals and Applications of Surface Phenomena Associated with Fouling and Cleaning in Food Processing* (ed. by B. Hallstrom, D. B. Lund and Ch. Tragardh) Lund University, 1981, p. 331.
78. Sanitary Design of Food Processing Equipment, *Food Processing*, 1964, Oct. and *Food Processing*, 1965, April.
79. Joint Technical Committee FMF/FMA, In *Hygienic Design and Operation of Food Plants* (ed. by R. Jowitt) Ellis Horwood, Chichester, 1980, p. 240.
80. British Standard 1501, Part 2, (1970), *Alloy steels for fired and unfired pressure vessels*, British Standards Institution.
81. British Standard 1501, Part 3, (1973), *Corrosion and heat resisting steel*, British Standards Institution.
82. British Standard 1470, (1972), *Aluminium and alloys*, British Standards Institution.
83. Code of Practice 3003, (1966), *Lining of vessels and equipment for chemical processes*, British Standards Institution.
84. British Standard 3719, (1979), *Stationary milk tanks (atmospheric type)*, British Standards Institution.
85. British Standard 3441, (1977), *Tanks for the transport of milk and milk products*, British Standards Institution.
86. British Standard 1864, (1966), *Stainless steel milk pipes and fittings (recessed ring joint type—RJT)*, British Standards Institution.
87. British Standard 3581, (1963), *Stainless steel cone joint pipe fittings*, British Standards Institution.
88. British Standard 4825, (1977), *Stainless steel pipes and fittings for the food industry*, British Standards Institution.
89. SMITH, G. A. and HEDRICK, T. I., *J. Milk Food Technol.*, 1967, **30**, 256.
90. MEYER, F., *Am. Soft Drink J.*, 1972, **127**, 22.
91. ASHER, H., *Dairy Ind.*, 1975, **40**, 341.
92. TIMPERLEY, D. A., *J. Soc. Dairy Technol.*, 1981, **34(1)**, 6.

Chapter 4

APPLICATION OF DIELECTRIC TECHNIQUES IN FOOD PRODUCTION AND PRESERVATION

MIHÁLY DEMECZKY

Central Food Research Institute, Budapest, Hungary

SUMMARY

In the chapter some possibilities of application of dielectric techniques in the food industry are outlined and a few important results gained in the above field are given.

The chapter is divided into five parts. In the first introductory part the role of water and the importance of the aqueous medium in the processing and preservation of raw materials in the food industry are discussed.

In part 2, the characterisation of dielectric heating as compared to conventional heating processes is to be found. The fundamental correlations of dielectric energy input and the importance of electrophysical properties of raw materials processed in the food industry are described. A brief survey of works carried out to determine the electrophysical properties of raw materials in the food industry is offered.

In part 3, the physical, chemical and biological effects induced by the use of high- and ultra high-frequency are dealt with.

In part 4, the possibilities of application of dielectric energy input in the field of the individual operations of the food industry with regard to the practical implementation and results of research works are discussed.

In part 5, the advantages of dielectric energy input in the food industry as against the conventional heating processes are briefly outlined.

1. INTRODUCTION

Modern physical processes are becoming increasingly applied in food production and preservation. A survey of the development history of technology shows that certain scientific discoveries or the application of general laws of Nature had significant effects upon the solutions of the practical problems of production. The acceleration of technical development in the twentieth century is facilitated by, among others, the rapid dissemination of the results of the various fields of sciences and the widespread application of these results in applied research. In the effort to ensure the well-being of mankind, those industries which immediately adopted scientific developments to their production technologies realised the most spectacular results. It is unnecessary to emphasise here the role of electrical energy in the development of modern industries, although the food industry was noticeably slow in adopting electronic techniques. This chapter deals with the application of dielectric techniques in the food industry and reviews the major results in the field.

Water is a necessary and very important constituent of the raw materials of the food industry. Water molecules, both free and bound, play a unique role in the structure of cells and the living material, in the formation of their microstructure and their physico-chemical characteristics. Both the primary and secondary processes of metabolism in plants and animals proceed, primarily, in an aqueous medium, and this aqueous medium plays an important role in the processing of the raw materials of food industry.

During the harvest, storage or processing of the raw materials of the food industry, physical, biochemical and microbiological processes can take place which reduce the quality of the end product—food. These processes can be eliminated, slowed or excluded by cooling, heat processing, reduction of the moisture content of the raw materials or by the deactivation of the catalysts active in these processes.

The raw materials, intermediates and end products of the food industry can be considered complex, porous colloid bodies with an intricate capillary structure. Water can be bound in these materials in extremely complex manners. The simplest means that can be used to characterise the state of hydration of these materials is their vapour pressure.

Most micro-organisms require for their growth a minimum amount of free or lightly bound water. Several studies have proved that micro-organisms fail to grow and multiply in materials in equilibrium with an air of 65% relative humidity. In the relative humidity range of 65–75%, only osmophilic yeasts show growth, while in the 75–95% range moulds grow as well. Above the 95% level bacteria also resume growth.

2. HEATING BY HIGH-FREQUENCY OR VERY-HIGH-FREQUENCY ELECTROMAGNETIC FIELDS

Heating is one of the fundamental processes of food preservation, both in sterilisation and pasteurisation, and in drying. The thermal conductivity of the raw materials of the food industry is, in general, low. Therefore, in the case of heat treatment of foodstuffs, there is a steep temperature gradient towards the interior of the material treated.

If the temperature difference is limited, then the time required for the processing is increased excessively. However, this increase in processing time is undesirable, because it leads to products of inferior quality. When the temperature difference is high, then the heating time is short, but certain parts of the foodstuff processed are subjected to excessively high temperatures and, once again, the quality of the product deteriorates. The unavoidable temperature gradient associated with conventional heating methods leads to an accumulation of moisture in the cooler parts of the solid material. The thermo-diffusion process involved increases in the inhomogeneity of moisture distribution in the solid material and limits the transfer of moisture towards the evaporating surfaces.

In principle, the previous undesirable processes can be avoided when dielectric heating is used. The construction blocks of the raw materials of the food industry are molecules, atoms and ions which, according to their size, electric charge and the alternating electromagnetic field, are forced to participate in a certain motion and absorb energy. The effects of high-frequency electromagnetic field are evenly distributed in the materials located between the electrodes or into the resonator cavity.

The power, P, absorbed by the material located in an electromagnetic field, is controlled by its dielectric loss coefficient as

$$P = 0 \cdot 56 f V (\varepsilon_r' \tan \delta) \left(\frac{U}{d}\right)^2 10^{-12} \text{ W}$$

where V = volume of the material (cm^3); f = frequency (s^{-1}); ε_r = relative dielectric constant of the material; δ = angle of the dielectric loss; $\varepsilon_r'' = \varepsilon_r' \tan \delta$ = dielectric loss coefficient; U = effective value of a sinusoidal (assumed) potential and d = distance between the electrodes (cm).

This observation offers a unique possibility for the high-frequency, dielectric excitation of water molecules, unevenly distributed in the solid raw materials of the food industry. This is so because the dielectric constant of water is some 20–30 times higher than that of the compounds which form the dry material of foodstuffs. The application of a high-frequency electromagnetic field permits, by the choice of the right conditions, natural or pre-dried materials with an inhomogeneous moisture distribution to enter the successive stages of processing at optimum temperature and moisture content.

In accordance with the natural structure of a foodstuff, its various parts absorb varying amounts of energy from the high-frequency field and, depending on the actual concentration of water molecules in that region, local temperature gradients can develop. However, these gradients do not follow a zonal arrangement, rather they are distributed over the entire mass of the material, according to the microstructure and the actual moisture distribution. Thus, the initial temperature differences lead to as rapid an elimination of the inhomogeneity of the moisture concentration as permitted by the thermal moisture–conductance process. The electrophysical properties of the raw materials and products of the food industry vary with their temperature, moisture content, chemical composition, physical structure, bulk density and the applied frequency.

The importance of the dielectric characteristics was recognised fairly early and substantial efforts were made to investigate the ε_r' and ε_r'' values of foodstuffs at various temperatures, moisture contents and frequencies. In the case of raw materials derived from plants or animals with chemical compositions and physical structures determined by genetic factors and the primary metabolic processes, the functions of the electrophysical characteristics vary with composition.[1,2] This permits understanding of the behaviour of materials during dielectric energy-transfer and permits the planning of proper dielectric energy-transfer processes.[3,4]

The variability of data published in the literature is readily explained, apart from the variability of the methods used, by the fact that materials of biological origin are involved. The development of methods to determine the dielectric characteristics can be followed by reviewing work with seeds. In 1953, Nelson et al.[5] described a method for the 1–50 MHz range, using the Boonton Q meter and a special sample holder. A publication from 1965[6] lists, apart from the definition of the dielectric characteristics, the ε_r', ε_r'' and conductance values of seeds at various moisture contents and frequencies in the 50 kHz– 50 MHz range. There were three publications in 1970, dealing with the electro-physical characteristics of seeds. They described the use of the Boonton RX Meter,[7] the General Radio Admittance Meter[8] and the General Radio 1608-A Impedance Bridge[9] and the special sample cells. A critical review of the electro-physical characteristics of agricultural products has been published[10] detailing the dielectric characteristics of various products at various frequencies, moisture contents, temperatures and volumes. In 1973, a paper describing a computer program for short circuited wave-guide dielectric properties measurements on high or low-loss materials appeared,[11] describing the principles and application of a computer program written in FORTRAN IV G. In the same year, another paper[12] dealt with the application of the coaxial method for the study of corn and weed seeds. Also in 1973, another paper[13] listed the dielectric constants and coefficients of dielectric loss for agricultural products, biological materials, foodstuffs, forestry products, feathers, rubber and soils, in the 10 kHz–10 GHz range. In 1974, two papers[14,15] discussed the use of computer programs for the precise calculation of the dielectric characteristics of low- and high-loss materials. In 1978 two papers[16,17] discussed the determination of the dielectric characteristics of corn at various temperatures, bulk densities, moisture contents and frequencies. In 1979 one paper[18] studied the dielectric characteristics of seeds and other foodstuffs, again at different temperatures, moisture contents, bulk densities and frequencies.

In 1979, a paper[19] showed an improved version of the cell used for dielectric measurements, carried out with a Q meter; a version suitable for a routine series of measurements. Another 1979 paper[20] showed, graphically, the dielectric characteristics of yellow corn plotted against its moisture content, temperature and bulk density. The paper described empirical equations which can be used for the calculation of the dielectric characteristics in the 20–300 MHz and 2·45 GHz ranges and which included such variables as moisture content, bulk density and temperature.

The dielectric characteristics of ground hickory nut in the 3–9%

moisture content range have been reported[21] at 12 different frequencies in the 50 kHz–12 GHz range. Both the dielectric constant and the loss coefficient increased with the moisture content and decreased with increasing frequency. The dielectric coefficient at a moisture content of 3% and frequency of 10 GHz was 1·8, while at 9% and 50 kHz its value was 13. The dielectric loss coefficient, under the same conditions, varied between 0·07 and 10·0.

A paper[22] published in 1982, discussed the major and minor factors which influence the dielectric characteristics of seeds. The most important factors include the moisture content and bulk density of the seeds and the frequency applied. Other factors, such as temperature, sorption and desorption characteristics, the chemical composition and structural heterogeneity are less important. The dielectric constant increases with the moisture content and decreases with increasing frequency. The loss coefficient and the tangent of the dielectric loss can both increase or decrease. Graphs show the values obtained for wheat and corn. A recent paper[23] discusses the density-dependence of the dielectric characteristics of various materials. It shows correlations between the square root and cube root of dielectric coefficient and density and between the loss coefficient and density.

Mynov et al.[24] measured the dielectric characteristics of pasta, enriched by 3–5% hydrolysed fish protein, in the 8–37% moisture content and 285–335 K temperature ranges. The tangent of the angle of dielectric loss and the dielectric coefficient, determined experimentally at various moisture contents and temperatures, allowed for the optimisation of the drying process of pasta using a combination of convective and high-frequency heating. Ponomareva et al.[25] studied the dipole moment of the phospholipids in sunflower seed and soybean oils. Both the hydrated and non-hydrated phospholipids showed high dipole moments in the 4·87–7·15 and 3·82–6·58 ranges. The dipole moment of sunflower seed oil was somewhat higher than that of the soybean oil. Rogov et al.[26] studied the dielectric characteristics of meat products at temperatures below 0°C. They determined the dielectric permeabilities and dielectric losses in the 183–263 K range.

Risman and Bengtsson[27] used the cavity perturbation method at 3 GHz frequency for the evaluation of the dielectric characteristics of a series of foodstuffs. Bengtsson and Risman[28] emphasised that a knowledge of the dielectric characteristics is important for the design of the heat treatment of foodstuffs, but added that there were other characteristics, equally important, influencing the actual heating of foodstuffs in microwave fields. Ohlson and Risman[29] showed that the design of the microwave oven, the pattern and distribution of load (standing waves), the extent of heat and material flow into the environment, the heat conductivity, enthalpy and density of foodstuffs all influence the optimum heating pattern.

According to Ohlsson,[30] the widespread use of dielectric, usually microwave, energy transfer to foodstuffs has been severely hindered by the lack of theoretical and design fundamentals, and development has been largely based on experimental results. Knowledge of the dielectric characteristics of foodstuffs, and their dependency on temperature, chemical composition and the frequency applied, is of prime importance. They found that data reported in the literature left much to be desired, so their work attempted to fill-in the gaps. They studied in detail five primary foodstuffs (meat, fish, roast beef/sauce, potato and water). They also extended the work, though briefly, to many other foodstuffs and food ingredients. They measured the dielectric characteristics in the 450–950 MHz range at temperatures from $-20°C$ to $140°C$, and at 2.8 GHz from $40°C$ to $140°C$. In order to investigate the role of composition they prepared 50 different meat emulsions composed of water, fat, protein and ash. Measured values were highly reproducible. They concluded that ε_r' and ε_r'' increased exponentially at the melting point temperature of the emulsions. In the 450–900 MHz range, ε_r'' increased rapidly with temperature. At 2.8 GHz, ε_r'' decreased with increasing temperature, up to $50°C$, then it increased. Penetration depths below $0°C$ showed surprisingly little variation. Calculations showed that the lower frequencies are more suitable for the thawing of deep-frozen foodstuffs, while higher frequencies are more suited for heating to higher temperatures because there the depth of penetration decreases less rapidly as the temperature is increased. A linear correlation ($r = 0.99$) existed between the ε_r' values and the water content of model meat-emulsions. The effects of the composition of meat emulsions upon the ε_r' and ε_r'' values can be shown readily in triangular diagrams, and conversely, composition can be calculated from known dielectric values. Computer programs have been developed for the calculation of the temperature distribution in the case of microwave heating, both above and below $0°C$. The calculations are based on the thermal characteristics of foodstuffs, on their dependency on the composition and temperature, on the dielectric penetration of energy and the laws of conventional heat conduction. These programs are used to evaluate the effects of the major factors which influence the rate of heating in thawing, direct cooking of frozen foodstuffs, in microwave sterilisation, etc. The factors included are the thickness and composition of the material to be heated, and the surface energy density and frequency of the field used.

The paper details the conclusions drawn from the computerised simulation studies as they apply to the heating, thawing, sterilisation and direct cooking of frozen foodstuffs, with special regard to HTST sterilisation. The paper emphasizes the importance of experiments in the verification of the computerised calculations.

Simovyan and Potapov[31] developed a mathematical model for the high-frequency heating of an infinite slab and for the dielectric heating of an infinite cylinder. The model serves as a starting point to ensure the homogeneity of high-frequency dielectric heating. Pfützner et al.[32] studied the dielectric characteristics of normal and PSE meats. They concluded that the phase angles of normal and PSE meats are different. Measurements were carried out by two blade electrodes 25 mm apart into the meat sample, parallel with the muscle fibres. Phase angles were determined for 675 muscle fibre samples obtained from 429 animal carcases. Characteristically PSE meat samples showed very low phase angles (between 1 and 5°). Since the phase angle of various muscle fibres, obtained from the same carcases, also differs, they proposed the use of the *gluteus medius* as a standard sample.

They concluded that the phase angle measured on normal meat samples was around 25°. Experience showed that when the phase angle measured on the *gluteus medius* was larger than 10°, then the other muscles of the carcase were free of the PSE problem, too.

Kleibel et al.[33] developed an instrument, based on the measurement of dielectric loss at 15 kHz, for the characterisation of the structure of meat. For normal meat samples the loss coefficient was about 2·5. For PSE samples the coefficient hovered around 40. Based on the comparison of the results of dielectric, free water, colour and fat ratio measurements carried out on 2000 meat samples they concluded that the dielectric loss, measured 15 min after the slaughtering of the animal, was a good indicator of the conditions of the meat sample 48 h later. The classification problems of PSE meats are also discussed.

There has been very intense work during the past 15 years on the evaluation of the dielectric characteristics of foodstuffs, agricultural products and other materials. Even a brief survey of this work is beyond the scope of this chapter. However, the papers of Nelson *et al.*[6-23] and the papers of Bengtsson, Ohlsson, Risman *et al.*,[27-30] as discussed above, indicate the extent and importance of this activity. It can be concluded that both ε_r' and tan δ vary continuously with the moisture content, bulk density, physical and chemical composition and structure of the material, with the temperature, and with the frequency of the applied field. This observation explains why it is more appropriate to carry out the dielectric treatment of materials in a continuous mode, via belts moving between set electrodes, rather than by loading the material, in a batchwise manner, into the cavity of the resonator. In the case of continuous operation the mass flow rate, average moisture content, average dry material content, average treatment time can be held constant, the integral value of the loss coefficient remains constant, and both the generator and the resonator can be operated under optimum conditions.

However, there is a significant temperature difference between the input and exit temperature of the material passing the electrodes. The temperature increases toward the exit point. On average, the amount of heat generated in the material is the highest at the exit point, as dictated by the ε_r' and tan δ values.

In materials with inhomogeneous dielectric characteristics the higher the loss coefficient (ε_r' tan δ) the higher the energy absorption and the amount of heat generated. If the material is to be heated homogeneously then the selection of the proper P/V value can eliminate the rapid rise in temperature of certain parts of the material. If certain parts in the material, with a tendency for rapid rise in temperature, cannot transfer the absorbed energy to other parts of the material—within the permitted temperature differences—then the rising temperature will lead to increased energy absorption, local overheating, breakdown of the insulator characteristics and rapidly decreasing energy absorption rate in other parts of the sample. The power level, at which these phenomena occur, is called the critical energy density. Under given conditions the value of the critical energy density is characteristic of the material. Thus, the inhomogeneity of the electrophysical characteristics of the material to be heated has to be taken into consideration when high-frequency fields are applied and the conditions have to be selected in such a manner that temperatures show a balanced rise and the energy density remain below the critical value. If the heterogeneous parts of the material are to be heated selectively, then the energy density has to be as close to the critical value as possible without, even occasionally, exceeding it.

The frequency of the field applied depends on the thickness of the

TABLE 1
DEPTH OF PENETRATION AS A FUNCTION OF
FREQUENCY

Frequency (MHz)	Depth of penetration (cm)
13·56	490
27·12	236
40·00	159
950·00	6·7
2450·00	2·6

Average for $\varepsilon_r' = 9$ and $\tan \delta = 0\cdot25$.

material, the homogeneity of power absorption in the treatment cavity and the value of the energy density to be realised. Depth of penetration and frequency are related (approximately) as

$$D = \frac{\lambda_0}{2\pi\sqrt{\varepsilon_r'}\,\tan\delta}$$

where D is the thickness of the material to be processed; λ_0 is the wavelength of the field selected; ε_r' is the relative dielectric constant of the material and $\tan\delta$ is the tangent of the loss angle of the material (Table 1).

3. THE EFFECTS OF HEATING BY HIGH-FREQUENCY FIELDS

The raw materials of the food industry are porous, capillaceous colloids. During conventional heating, drying or moisture tempering processes their structure changes and this change is accompanied by shrinking. The unavoidable high temperature gradients involved in conventional thermal processing influence the mechanical behaviour of foodstuffs very unfavourably and render their structure in-homogeneous. This effect is manifest even in the intermediate phases of processing, especially when large amounts of water have to be evaporated. There is a fairly intense transfer of moisture from the outer layers of the material toward the surface and it is accompanied, due to the law of thermal diffusion, by a transfer of moisture towards the interior of the material. Thus, a crust of low moisture content and heat conductivity is formed, preventing the moisture trapped in the core of the material from reaching the surface and slowing the flow of heat from the surface toward the core of the material. This process is also accompanied, in most cases, by a change in the microstructure of the material.

If the moisture content of the material is reduced by the application of microwave fields then, following the initial period of equalisation of the inhomogeneities of moisture content, no significant gradients occur in either the temperature or the moisture content. The moisture content and temperature of the material can be considered homogeneous, within narrow limits, and induced heat generated all across the material, homogeneously turns the moisture content of the material into steam. Liquid contained in the pores and capillaries is exuded. In the next stage, liquid trapped in parts of the material with no direct access to the surface or to the larger cavities is turned to steam at a rate dictated by the rate of energy absorption. When most of the material has a direct connection with the surface, then quite large amounts of liquid can be removed with a very small amount of energy. However, when there is no direct conductance to the surface, or when conductance is restricted, then the moisture is turned to steam at a rate determined by the rate of power uptake.

Accordingly, the amount of power absorbed by a unit of mass of the material in a unit of time determines the eventual microstructure of the material. When the rate of energy input is optimised, then the dried, heated material does not shrink. In fact, if the rate of energy input is fairly high, then the material can expand resulting in the opening up of the original structure and the formation of a new microstructure. During the entire cycle of drying, energy can be transferred uniformly over the entire cross section of the material. Because of the development of the favourable microstructure and the exclusion of the process of thermodiffusion, the energy consumption can be much lower than in the case of conventional thermal processes. The opening of the microstructure changes the hygroscopic characteristics of the material, especially in the case of colloidal–capillaceous materials. This often reduces the extent of capillary condensation under the usual relative humidities encountered in storage. Since the amount of energy and its input rate can be controlled quite precisely in the case of high-frequency fields, it offers a unique and reproducible way to produce well-defined microstructures. It is unnecessary to say how important these well-defined microstructures are during subsequent pressing, extraction and preservative addition.

Chemical processes, commonly encountered in the heat processing of foodstuffs, occur in dielectric processing too. Apart from the occasional hindrance of catalytic processes, these reactions proceed faster at elevated temperatures. In fact, some experimental evidence shows that the rate increases realised are higher than those calculated from Arrhenius' law. In certain cases one has to assume that some reactions become exceedingly activated in high-frequency fields. In the case of more complex processing steps, for example in fermentation or ripening where microbial, enzymic and chemical processes occur simultaneously, the enzymic and/or microbial processes can be excluded selectively. Experimental evidence shows that foodstuffs sterilised by dielectric processing retain their original chemical composition better than those sterilised by conventional methods.

Certain characteristics of certain intermediates or end products of the food industry can often be improved by the application of high-frequency fields. For example, the acid value of pasta can be reduced by dielectric treatment. Regarding the change of the dry-material content of certain foodstuffs during processing it can be concluded that, compared with conventional technologies, the extent of dry material loss during dielectric processing is much lower, often a fraction of what is encountered in conventional technologies. The decrease in the change of the dry material content in dielectric processing is in agreement with the observation that the oxygen index of dielectrically processed materials decreases to a fraction of the value measured before the processing step. Such drastic reductions with conventional technologies would require much longer fermentation or storage periods.

Studies dealing with the effects of dielectric processing upon enzymes are still in their infancy, as are chemical investigations. Several workers have studied the effects of conventional and dielectric heating, under comparable conditions, and found neither special activating nor deactivating effects.

Peroxidases, catalases and polyphenoloxidases can readily be deactivated by dielectric treatment. This observation is of great importance in the storage of certain materials. The characteristic effects of dielectric treatment are especially important in reducing the duration of time-consuming processing steps. As mentioned above in connection with the changes of the dry material content and the oxygen index, dielectric processing is not accompanied by such extensive changes in chemical composition as observed in conventional technologies, which require much longer processing times. Favourable and more extensive microstructure-modifying processes occur in dielectric processing than in conventional processing.

The effects of high-frequency and very-high-frequency fields upon the life of micro-organisms have been studied extensively during the past 50 years. These studies were directed, first of all, at proving or disproving the special effects upon the micro-organisms of high- and very-high-frequency fields. Several bacteria and fungi were tested, under optimum conditions, using both conventional heat transfer techniques and high-frequency and very-high-frequency fields alike. In other tests, the temperature of test culture and control culture was increased at the same rate by both conventional and dielectric means of heating, up to the value required for sterilisation.

The selective effects of high-frequency and very-high-frequency fields upon the micro-organisms have been debated for years. It can now be concluded that the primary effects are due to the heat induced in the micro-organisms and effects peculiar to high-frequency treatment are but of secondary importance. Considering the large number of variables (frequency, field strength, nutrient solution, rate of energy input, actual biological state of the micro-organisms), no final conclusion can be drawn, even though the number of experiments completed is considerable. Wide-ranging studies in food microbiology over the past years indicate that the sterilizing effect of dielectric and microwave heating is due, primarily, to their thermal effects. This conclusion can be reconciled with those apparent contradictions which state that in high-frequency and very-high-frequency fields certain micro-organisms can be sterilized at comparatively low temperatures. Considering the different electro-physical characteristics of various micro-organisms it can be argued that colonies of the micro-organisms selectively absorb energy and their actual temperature might be, locally, quite different from that of their surrounding.

High-frequency and very-high-frequency treatment offers definite advantages for sterilisation, related to the rapid rise in temperature. When valuable biological materials are to be preserved, pasteurisation or sterilisation by microwave heating is especially advantageous. The time required for sterilisation is several orders of magnitude shorter than in the case of classical heat sterilisation, because the slow process of heat penetration is absent. This feature can be utilised in the development of continuous processing technologies.

The inhomogeneous distribution of moisture, especially in plant-derived materials, serves as a self-control mechanism and extends the duration of energy absorption from the high-frequency field. Parts of the material with a higher initial moisture content reach higher temperatures in the first stage of the process. This significantly increases the storage life of such inhomogeneous materials.

The precise control during processing of both the temperature and moisture content of the raw materials of food industry is of fundamental importance in each step of the technology. It has been mentioned that homogeneous heating can be realised with dielectric energy input and, except for small differences due to the structural and/or chemical characteristics of the material, no zones of unequal temperature have to be reckoned with. The fact that the energy uptake of the material is proportional to its moisture content in the initial stage of heating and that local, temporary temperature gradients tend to create a uniform moisture content in the material make the control of this moisture content in the successive stages of production simple. The relationship which describes the energy absorption of insulators and semi-conductors in a high-frequency field indicates that the rate of energy input can be precisely controlled via the field strength and the frequency of the electromagnetic radiation. A third factor, time, can also serve as control variable. The high-frequency field can be applied intermittently, for fractions of a second, or continuously as long as required. This permits the adjustment of the temperature best suited to the selected process. The time required to reach a certain temperature can also be controlled and it can be orders of magnitude shorter than in the case of radiation, conduction or convection heating. This has the added advantage that the favourable temperatures for certain microbial processes occur only for very short periods of time and, consequently, these processes remain insignificant.

Having reviewed the physical, chemical and biological effects that can be realised by the application of high-frequency and very-high-frequency fields we will consider now, in detail, the results of laboratory and plant-scale experiments pertaining to the various branches of the food industry.

4. APPLICATIONS OF DIELECTRIC HEATING

In the various branches of the food industry, dielectric heating can be applied for (1) heating, drying, roasting; (2) preservation, sterilisation; (3) pre-cooking, cooking, baking; and (4) thawing of frozen products.

4.1. Heating, Drying and Roasting

Demeczky *et al.*[34,35] have developed an industrial-scale, continuously operated, high-frequency fat-melting apparatus which also produces non-denaturated protein. The advantages of homogeneous energy input and precise control are utilised to melt animal fats with very low thermal conductivities.

Compared to the conventional melting pots, the time required for melting is decreased 10-fold in the dielectric process. The precise control possibilities facilitate the setting of the optimum temperature for the given type of fat and avoid the denaturisation of protein fibres. The lack of temperature gradients leads to two separate phases, to lard and protein fibre, so all the disadvantages of both the dry and wet fat-melting processes can be avoided.

The schematic of an industrial scale, continuously operated, 1·2 t fat/h capacity system, equipped with a 25 kW, 13·56 MHz generator, is shown in Fig. 1. The technical and economic characteristics of both the conventional and dielectric process are shown in Table 2.

Yamaguchi[36] reported the construction of a 13·5 kW, 2·45 GHz microwave heater used for the production of kamboko gel. Gels with 50 and 80–100% added water could be produced by 0·017 and 0·03 W h/g power applied for 10·4 and 18·5 s/100 g, respectively. Compared to the conventional systems the energy cost was 44·5% and the amortisation time of the system was 1·23 year.

Caddeda[37] described a microwave unit used for the production of Mozzarella cheese. The conventional method offers a lower level of fat-loss and allows direct salt application, but microwave heating is

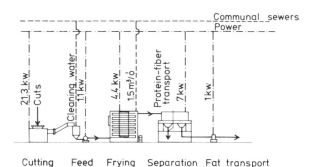

FIG. 1. Technological schematic of a continuously-operated high-frequency fat-frying unit with protein recovery.

TABLE 2
COMPARISON OF PRODUCTION COSTS IN THE CASE OF DRY-FRYING IN
DUPLICATORS AND HIGH-FREQUENCY FRYING LEADING TO NON-
DENATURATED PROTEIN FIBRES (IN THOUSAND FORINTS, FOR AN
ANNUAL PROCESSING CAPACITY OF 4400 t/year AND YIELD OF 1 t/h)

Name	Dry-frying in duplicators	High-frequency frying
Raw material to be processed	72 600	72 600
Processing cost	610	563
Net direct cost	73 210	73 163
Value of recovered product	74 100	103 286
Direct technological profit	890	30 123

better for the thawing and heating of frozen products. Heating by microwave along a conveyor belt is especially advantageous for the equalisation of the daily fluctuations in consumption.

Entremant and Levardan[38] described a melting process and apparatus for the separation of the fat content and other products of butter. Melted butter is fed into a microwave system. Butter flows through a channel with a U cross-section and it is subjected to a microwave field. This treatment separates fat, protein and water, and the temperature rise is kept to a minimum. Tests showed that the efficiency of separation was 99·5% and that temperature rise during the 1 min long irradiation step was only 6°C. The product is dried to a dry-material content of 99·5%, it is pasteurised and packaged.

As an example for the heating of heat-sensitive blocks we should mention the heating of raw chocolate blocks of 25–30 kg weight to temperatures of 30–35°C. Using conventional techniques this warming process is a very slow and delicate one, because even minor local overheating may lead to an entirely undesirable taste for the whole block. During warming, temperature has to be controlled to at least 0·1°C and this requires the application of a very time-consuming uniform heating process. Using high-frequency fields this heating task can be realised very precisely and easily. Now chocolate of 1500–2000 kg can be warmed with the help of a 50 kW nominal capacity generator.

Pour-El et al.[39] studied the chemical and biochemical changes which occur during the dielectric heating at 42 and 2450 MHz frequencies of whole soybeans as a function of the amount of absorbed energy. Since the rate of temperature rise was different at the two different frequencies, the time-dependency of the various biochemical characteristics also appeared frequency-dependent. However, when the results were plotted against the amount of absorbed energy, the apparent frequency-dependency disappeared. The chemical and biochemical tests showed that dielectric heating of soybeans of natural moisture levels was as effective as conventional steam toasting in increasing the nutritional value of soybeans. Trypsin inactivation proceeded to such an extent that maximum nutritional values could be realised. Protein solubility and dispersibility were used to follow the extent of trypsin deactivation during dielectric heating. Tests based on the measurement of the urease activity level proved unsuitable for this purpose. The lipoxygenase activity could be completely deactivated by dielectric treatment, though with certain samples a slight residual activity was observed. This might be advantageous, together with a high peroxidase activity, for decolouring.

Demeczky et al.[40] described the high-frequency treatment of sunflower seed. Neither the peroxide value nor the acid number of the treated seeds changed during extended storage. Consequently, the development of rancid taste can be avoided. The microstructure achieved at 70°C is favourable for the recovery of oil. The yield of oil in cold presses is higher, and higher pressures can be applied. Results of an industrial-scale pressing test are shown in Table 3. The quality of oil is better and it can be raffinated more easily. A paper[41] discusses the high-frequency treatment of rapeseed. The extraction of rapeseed

TABLE 3

RESULTS OF OIL-PRESSING EXPERIMENTS CARRIED OUT WITH DIELECTRICALLY TREATED AND NON-TREATED, DEHUSKED, FULL SUNFLOWER SEEDS

Method of treatment	Oil yield	Residual oil content in press cake
Control seed heated to 72–95°C by indirect steam	36·5	24·4
Seed dielectrically heated to 72–80°C	43·6	14·2

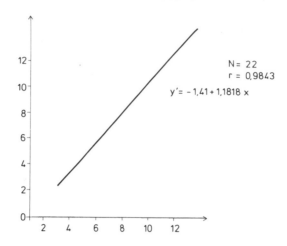

Fig. 2. Improvement in % wt/wt oil extraction (y') upon heat treatment in high-frequency field, related to the overall oil content of seed, plotted against the residual oil content of extracted seed without dielectric treatment (x).

(obtained from different areas and years) proved more complete when the seeds were subjected to high-frequency treatment prior to the extraction step. Seeds, treated at 105–115°C by a power level of 0·3–1·0 W/g, could be extracted more completely, even if stored for several more weeks, than seeds treated by conventional methods. The results are shown in Fig. 2. Mahesvari[42] studied the effects of microwave treatment upon rapeseed samples. *Brassica napus* cv. Tower, *Brassica compestris* cv. Candle and *Brassica juncea* oriental mustard seeds were tested. He found that treatment of the dehusked seeds led to the inactivation of thioglucoside glucohydrolase. The effects of this deactivation upon the quality of oil and crushed rapeseed were also studied.

Pedenko *et al.*[43] reviewed the changes of the quality indicators of fat constituents during microwave heating of meat. Conventional heating techniques were applied for comparison. Soxhlet extraction was used to recover the fat content of meat samples. The acid value, peroxide value, iodine value, thiobarbituric acid value, free radical concentration and fatty acid composition were also determined. They concluded that microwave heating was suitable for the processing of raw meat into meat products. No harmful effects were recorded.

Drying curves recorded during the drying of very moist materials of biological origin indicate that some 80–85% of the moisture content leaves the sample during the first 50–60% of the drying period. The removal of the last portion of moisture by air drying requires a disproportionally long drying time and too much energy. Dielectric methods can be used advantageously for the removal of the moisture content at the flat portion of the drying curve. Such a combination of conventional and dielectric drying methods does indeed decrease the specific energy consumption of drying and increases the actual production capacity of existing driers. Demeczky and Gondár[44] studied the high-frequency air drying of various tobacco products. They concluded that the catalase value decreased, the storage characteristics improved and the hygienic requirements could be met more readily. Demeczky[45] developed a method for the final drying of hops using a combination of conventional and high-frequency drying techniques. The method was used to adjust the homogeneous moisture content of hop bales required for storage. The technological characteristics of hops, the specific energy consumption and drying time requirements were compared for the conventional and the combined drying techniques. The electrophysical characteristics of bulk hops were also determined. A pilot-plant-scale, continuously operated, dielectrically-heated drying belt is also described. Laboratory measurements were carried out monthly to follow changes in the total resin content, hexane-soluble resin content, alpha-acid, beta-acid, gamma-resin and bitter value of hop batches subjected to different drying techniques, for a storage period of 1 year. Thirty-seven plant-scale beer brewing experiments were carried out. The average values are listed against the brewing time in Table 4. The results show that specific energy consumption can be decreased by 20–30% in the case of combined drying, while the actual drying capacity of the conventional drier can be increased by 50–70% with the addition of the dielectric air-drier. The brewing time of hops can be reduced from 90 min to 50 min. Hops dried by the combined method retain their bitter value better; consequently, some 10–20% of hops can be saved.

The bottleneck in dried pasta production lines is the end drier section. Demeczky[46] studied the combination possibilities of conventional and dielectric drying. He determined the ε_r' and tan δ values of various pasta compositions for moisture contents (6–22%) and temperatures (20–50°C) encountered in actual pasta drying. The high-frequency drying curves of the most popular pasta products (short-cut macaroni, spaghetti, vermicelli, straight noodles, large elbows, grains) were recorded at atmospheric and reduced pressures. Based on the

TABLE 4
AVERAGE BITTER VALUES AND AL NUMBERS OBTAINED
FOR VARIOUS BREWING PERIODS

	Brewing time (min)	Bitter value	AL number
Average value relating to dielectrically dried hop	15	29·6	70·2
	20	30·6	72·1
	30	32·6	66·0
	50	39·5	70·0
	90	38·2	70·0
Average value relating to conventionally dried hop	90	31·1	69·7

drying curves of a conventional, plant-scale, end-drying belt and the dielectric drier operated at atmospheric pressure, he designed the drying curve of a combined plant which relies on both the conventional and the dielectric modes of drying (Fig. 3). When the actual drying curve of the combined plant was recorded, drying times much shorter than expected were observed (Fig. 4). The cooking characteristics of pasta products dried under different conditions were also studied. It was concluded that temperatures in excess of 40°C should not be applied because this increased the cooking loss. This conclusion, however, does not apply to the production at reduced pressure of

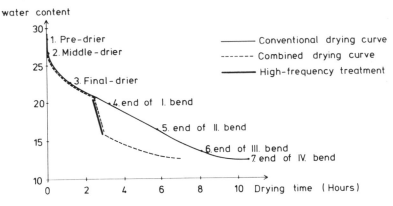

FIG. 3. The proposed combined drying curve of short cut macaroni (industrial-scale unit).

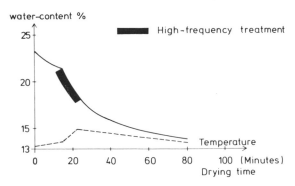

Fɪɢ. 4. Combined drying curve of short cut macaroni (on the basis of research data).

low-moisture-content pasta products (at and below 6%), where final drying temperatures as high as 80°C could be applied without harmful effects. It was concluded from the experimental results that the production capacity of conventional, continuously operated pasta-producing lines could be increased by at least 25% by the addition of a dielectric end-drier to reduce the moisture content of the product from 20% to 12·5%. No extra construction work is needed for this addition and compared with conventional production capacity boosting methods the capital investment costs are much lower. The quality of pasta produced by the combined technology is identical or better than that obtained on the conventional line.

A good example for the rapid removal of low moisture contents is the dielectric drying of cube sugar.[47] Sugar is mixed with 1–2% water and rods are pressed, then the rods are cut to cube size. The added moisture content of cube sugar has to be removed, though. This task can be elegantly solved by dielectric heating on a continuously operated conveyer belt. While a decrease of the initial moisture content to 0·2–0·25% by the most modern infrared drying tunnel requires 64·5 kW h/t, experiments carried out in the Central Food Research Institute showed that by dielectric heating the same task could be achieved with one-third of this energy.

Akahashi and Matashige[48] carried out modelling studies of microwave air-drying. They used moist 20 mm thick, woven paper as model material. Drying was accomplished by flowing air in a microwave system. The temperature of the inner layer of paper was maintained at a pre-set level by controlling the amount of energy introduced by the microwave field. The distribution of moisture in the model material depended on the temperature difference between the inner layer and the air flow. When this temperature difference was less than 20°C, the moisture content of the inner layer was less than that of the outer layer. When the temperature difference was larger than 30°C, then the outer layer was more moist. At a temperature difference of about 25°C the moisture distribution was homogeneous. The analysis of the data suggested that foodstuffs can be dried evenly by selecting an appropriate temperature difference. It was found that the rate of drying decreased proportionally to the velocity of the heat-dissipating air flow when the temperature of the drying air was lower than that of the dried sample.

Kamai *et al.*[49] studied the microwave drying of onions and carrots. Microwave drying was compared with conventional freeze drying, hot air drying and vacuum drying. The level of water-absorption was high for freeze-dried samples and low for microwave-dried samples. Samples subjected to freeze drying were softer than those subjected to microwave drying. The colour of the dried products depended on the temperature of drying. The colour of microwave-dried samples was similar to that of the samples dried at 80°C by flowing air. The colour of freeze-dried samples matched the colour of air-dried samples which were produced at 60°C. In the case of carrots, the drying method used had almost no effect upon the colour of samples, but the blanching method used did influence the final colour. Electroscanning studies showed that the texture of microwave-dried samples shrank more than that of the air-dried samples, for both onions and carrots.

Leonhardt and Gomez[50] studied the microwave drying of the biomass of *Saccharomyces cerevisiae*. The effects of particle size (0·57, 1·42 and 2·68 mm), layer thickness (4, 8, 12 and 16 mm) and power input (1·2, 1·4, 1·6, 1·8, 2·1, 2·4, 3·0 and 3·6 kW) were studied. Mathematical analysis of the numerically and graphically presented data showed that the particle size had no effect upon the drying operation while increased energy input decreased the duration of the drying period. Specific energy consumption improved as the layer thickness increased. Mathematical relationships were derived allowing for the design of proper drying schemes.

Fregni and Leonhardt[51] studied the drying of soybeans with a moisture content of 14·3–20·0% in a cubic microwave oven. Input energy was varied between 0·2 and 2·4 kW. The amount of both the absorbed and reflected energy, the initial, intermediate and final moisture contents were all measured. Estimates were made as to the expected energy consumption of a plant-scale drying unit. The energy consumption of the experimental unit was rather high. They discussed the possibilities of energy optimisation in the case of the plant-scale unit.

Demeczky and Márkus[52] studied the high-frequency dielectric drying of low moisture content pastas produced at reduced pressure. It is essential that the vapour pressure of pasta products used in soup-powders and other low moisture level products be low. The moisture content of such pastas has to be kept in the 3·5–4·5% range. The electrophysical characteristics of moisture-lean pastas were determined in the 3·5–12% moisture content range and in the 20–100°C temperature range. Under the experimental conditions used corona-discharge phenomena could be avoided at a pressure of $1·03 \times 10^4$ Pa. They concluded that the comparatively high final temperature of drying, 80°C, has no deleterious effects upon the cooking characteristics of pastas. The drying times varied, depending on the type of pasta, between 100 and 200 min. Preliminary calculations showed that the plant could be realised economically.

Brighenti et al.[53] studied the microwave vacuum drying of egg white, horse meat and veal. They found that microwave drying eliminated 95% of the pathogenic bacteria, had no deleterious effects upon the SH functional groups, modified the cellular structure only slightly, did not cause denaturation and did not hamper the emulsification characteristics.

Meisel[54] described a combined microwave-heating, infra red-sensing drying system. Its plant-scale version is available commercially. Drying is carried out at a pressure of 0·67–1·33 kPa, using a nominal 6 kW magnetron controlled by six conventional and two infra red sensors. The system is called the Gigavac. Fruit concentrates or solid foodstuffs are moved by a conveyor belt and drying at 40°C takes about 60 min. The powder or particulate product has an instant nature and its quality is much better than that of the comparable freeze-dried product. Concentrates of 60° Brix can also be dried in the system. Economic calculations show a very favourable picture: the drying costs of 1 kg end product are 1/3 of the cost of the comparable freeze-dried product. Ang et al.[55] carried out freeze drying experiments with frozen beef using a field strength of 100, 130, 150 and 170 V/cm and a pressure of 33·25, 79·99, 126·66 and 199·98 Pa pressure. The moisture and ice content of the samples were determined at various drying times. They concluded that under the experimental conditions used (2450 MHz) the field strength of 130 V/cm was optimum. Higher field strengths resulted in the overheating of the dried products. Drying took 4·33 h at 130 V/cm and 33·25 kPa pressure. In freeze drying, pressure has to be kept as low as possible. They found that pressures below 26·66 kPa failed to reduce the length of the drying period. Arsem et al.[56] described a microwave system to be used in the freeze drying of meat. A unit of 7 kg capacity, employing microwave and radiation heating, was used. The system allowed much shorter drying periods than the comparable systems which used radiation only. Sublimation was faster than expected because sublimation of ice required less energy than calculated. It seems possible that the vapour flow carried away partially sublimed particles, too. They note that these possibilities should be contemplated in the design of freeze drying systems. Sunderland[57] reviewed the present state-of-the-art of microwave freeze drying. He presented a general evaluation and put forth propositions for future research. He believes that the major obstacles to microwave freeze drying have been eliminated. Economic studies indicate that microwave freeze drying is cheaper than conventional freeze drying. He claims that the results of experiments and development projects published in the literature form an adequate basis for the design of industrial systems. Practical advice and hints for the realisation of industrial systems are also given. Proposals are advanced as to the desirable future course of research and development. Sunderland[58] discussed in detail the economics of microwave freeze drying and compared it with conventional freeze drying. He concluded that the current overall cost of conventional freeze drying is 0·279 US $/kg while that of the microwave process is only 0·185 US $/kg.

Roasting plays a fundamental role in the processing of most oily seeds. Part of the changes which take place during roasting can be followed by laboratory measurements (moisture content, fat content, dry material content, chemical composition), but detection of the characteristic flavour, aroma or scent components is done, mostly, by sensory evaluation, though there is active work aimed at the introduction of instrumental analytical methods for the purposes of aroma characterisation. Flavour and aroma depend on the temperature and duration of roasting. The uniform application and strict maintenance of roasting temperature and time, determined by sensory evaluation, can be realised by the application of dielectric heating. Demeczky[59] studied the high-frequency, dielectric roasting of cocoa beans. Conventional and high-frequency roasting techniques were used and the dry material content, husk and core moisture content and cocoa butter content were compared (Table 5). The major electrophysical characteristics of cocoa beans were also determined at various temperatures, at a moisture content of 3%. The operating conditions of continuous dielectric roasting of cocoa beans in a plant-scale unit were also evaluated. Test roastings in a pilot-scale unit (Fig. 5) showed that the loss of dry material was 1·25% lower in the case of dielectric roasting. Since the fat loss of the roasted husk also decreased, a cocoa butter loss of 0·2% could be avoided. High-frequency roasting significantly improved the separation characteristics of husk and core. The economics of high-frequency roasting of cocoa beans is detailed in Table 6. Kuzhnetsova et al.[60] studied the variations in the soluble tannin content, fat content, fat content and ammoniacal nitrogen content of the reducing materials of cocoa beans during high-frequency treating applied for a period of 180–540 s. The composition of high-frequency treated cocoa beans was compared with that of a conventionally roasted cocoa bean sample (30 min roasting time). Under the conditions used the composition of cocoa beans treated for 420–480 s proved the best, while beans treated for 540 s had a burnt smell. The reducing material content, ammonia nitrogen content and soluble tannin content of high-frequency treated and conventionally roasted cocoa beans were similar. The costs of high-frequency roasting were as low as 20–25% of the conventional process, and homogeneous heating in the high-frequency field had a favourable effect upon the quality of the product.

Krüger and Groenick[61] described the fundamentals of the dielectric treatment of foodstuffs and reported on the laboratory-scale drying of malt. They designed an experimental microwave malt drier of 100 g green malt capacity. Malt was dried by 60 W for 4 h, then by 90 W for 1 h. The moisture content, extractables content, alpha- and beta-amylase activity and nitrosamine concentration of the conventionally dried and microwave dried samples were compared. The EBC (colour) value, Kolbach's value, Hartong's value, extract pH and aroma intensity were determined. The results show that high-quality malt could be obtained by dielectric drying. The energy balance and the economics of the process are also discussed. Luter et al.[62] studied the microwave decomposition of the aflatoxin content of peanuts. They found that the extent of decomposition was independent of the rate of

TABLE 5
RESULTS OF COCOA BEAN ROASTING BY CONVENTIONAL AND HIGH-FREQUENCY METHODS

	Power density (W/g)	Time of treatment (min)	Final roasting temperature (°C)	Moisture content of beans (%)	Moisture content of husk (%)	Fat content of husk (%)	Roasting loss (%)
High frequency roasting	0·55	9	142	2·4	5·0	2·07	3·6
	0·60	8	143	2·2	4·6	2·18	3·5
	0·56	9	145	2·1	4·7	2·11	3·9
	0·52	10	146	2·1	4·3	2·20	4·0
	0·58	8	144	2·3	4·2	2·27	3·9
	0·55	9	146	2·1	3·8	2·05	3·9
	0·53	9	144	2·3	4·1	2·30	3·7
Conventional roasting	—	23·8	147·4	1·78	3·02	5·2	6·1
Raw material	—	—	—	5·1	12·7	1·79	—

FIG. 5. Cocoa bean roaster unit.

TABLE 6

THE ECONOMICS OF HIGH-FREQUENCY ROASTING

	Roasting line Sirokko 4002	High-frequency
Annual raw beans consumption (t)	2600	2·660
Annual roasted beans production (t)	2·250	2·278
Roasting and dehusking losses (t)	410	382
Roasting and dehusking losses (%)	15·4	14·4
Value of roasted beans (51 800 Ft/t Thousand Ft)	116·550	118·000
Difference in value of roasted beans (Thousand Ft)	—	1·450
Direct plant costs of roasting (Thousand Ft)	337	615
Excess costs of high-frequency roasting (Thousand Ft)	—	278
Direct technological extra yield of high-frequency roasting year/line (Thousand Ft)	—	1·172

Roasted bean is defined as dehusked, roasted, ground material.
Ft = Hungarian forint.

heating. Heating peanuts by microwave up to 150°C decomposed some 95% of its aflatoxin content. The extent of browning was adequate while proteins and fatty acids were not affected. Gerling and O'Meara[63] obtained a patent for a microwave coffee roasting apparatus. According to the patent, coffee, flowing through a transparent drum, is roasted continuously by microwave heating.

4.2. Preservation and Sterilisation

Demeczky[64] described a continuously operated high-frequency fruit juice pasteurising system, developed by the Technology Department of the Central Food Research Institute. Along with the advantages of packaging with plastics, rapid heating and cooking can also be realised by the system. It can be used for the rapid, mild and economic heat treatment of fruit juices and vegetable juices or diluted concentrates. The products obtained comply with modern quality requirements. Polypropylene flasks, moved by gravity, proceed through the apparatus which has been designed on the basis of the electrophysical characteristics of fruit juices and vegetable juices and similar model solutions, in such a manner that intense mixing is also ensured. The schematic of the pilot-plant scale unit is shown in Fig. 6. The aroma

FIG. 6. Schematic of the electrode system of a continuously operated experimental high-frequency pasteurising unit.

TABLE 7

THE AROMA VALUE OF FRESH SOUR CHERRY JUICE DURING CONVENTIONAL AND
HIGH-FREQUENCY PASTEURISATION

	Fresh sour cherry juice	Conventional pasteurisation	High-frequency pasteurisation
Aroma value (oral test results)	30·18	22·80	29·20
	30·18	23·11	29·60
	30·38	22·53	29·75
	30·38	22·34	29·60
	30·18	22·51	29·60
	30·18	22·51	29·75
Average aroma value	30·25	22·63	29·58
Standard deviation	0·34	1·22	0·68

values of the conventionally and dielectrically pasteurised products are compared in Table 7, while the overall micro-organism counts are shown in Table 8. Economic calculations showed that the cost of dielectric pasteurisation was only 60% of the cost of the conventional process.

Bookwalter et al.[65] studied the effects of microwave treatment in corn-milk and soymilk mixtures with respect to product quality and residual *Salmonella* activity. Enriched corn-milk was mixed with *Salmonella senffemberg* and was packaged into units of 22·7 kg.

TABLE 8

OVERALL COUNT IN FRESH AND PASTEURISED SOUR CHERRY JUICE

Samples	1	2	3	4	5	6
Overall count (10^6/cm^3)	28·5	13·6	14·1	14·0	14·9	14·0
Live yeast cells						
Noble yeast/cm^3	1·83	1·24	—	—	—	—
Wild yeast/cm^3	48	42	—	—	1	0·3
Mould/cm^3	3	2	—	—	—	—

(1) Fresh, non-pasteurised
(2) Fresh, non-pasteurised
(3) High-frequency pasteurised
(4) High-frequency pasteurised
(5) Conventionally pasteurised
(6) Conventionally pasteurised

Packages with both natural and artificial *Salmonella* contents were
heated, in 3·9–10·0 min, from 21°C to 82·2°C in a continuously
operated, 60 kW rated (2450 MHz) dielectric tunnel. Starting with an
initial *Salmonella* count of $4 \times 10^2\,g^{-1}$, the *Salmonella* concentration
could be reduced by 10^2 by heating to 56·7°C, and by 10^5 by heating to
82·2°C. At 82·2°C the nutritional value of the product decreased, but
no such decrease could be detected either at 61·1°C or 67·2°C.
Demeczky and Márkus[66] studied the effects of high-frequency treatment upon the artificially contaminated (with *Staphylococcus aureus*
and *Salmonella*) pasta products. The electrophysical characteristics of
the various pasta products were determined in the 50–100°C range.
Samples were heated in paper bags to final temperatures of 60, 80 and
100°C, and the microbiological and cooking tests of the cooled samples
(25°C) were completed. The results of the microbiological tests are
shown in Tables 9 and 10, while the cooking tests are detailed in Table
11. At a final temperature of 100°C all the *Salmonella* activity could be
eliminated.

Nosikov *et al.*[67] studied the sterilisation in high-frequency fields of

TABLE 9

RESULTS OF THE MICROBIOLOGICAL ANALYSES OF A PASTRY SAMPLE INFECTED
BY *S. aureus*

	Staphylococcus count/g		*Final temperature of heating (°C)*
	Without high frequency	*With treatment*	
Shells	$4\cdot2 \times 10^4$	9×10^1	60
	$4\cdot2 \times 10^4$	9×10^1	80
	$4\cdot2 \times 10^4$	9×10^1	100
Square-flake	$5\cdot3 \times 10^4$	9×10^1	80
	$3\cdot4 \times 10^5$	9×10^1	80
	$1\cdot3 \times 10^4$	9×10^1	80
	$1\cdot4 \times 10^4$	9×10^1	80
	$1\cdot2 \times 10^4$	9×10^1	80
	$2\cdot1 \times 10^4$	9×10^1	80
	$1\cdot4 \times 10^4$	9×10^1	80
	$1\cdot8 \times 10^4$	9×10^1	80
	$1\cdot4 \times 10^4$	9×10^1	80
	$2\cdot0 \times 10^4$	9×10^1	80
Square-flake small	$5\cdot1 \times 10^4$	9×10^1	85
Grains	$7\cdot0 \times 10^5$	$2\cdot3 \times 10^3$	60

TABLE 10
RESULTS OF THE MICROBIOLOGICAL ANALYSIS OF DRIED GRAINS—
PRODUCTS ARTIFICIALLY INFECTED WITH *S. aureus*

Staphylococcus count/g		Final temperature of heating (°C)
Without high frequency	With treatment	
$2{\cdot}0 \times 10^7$	$4{\cdot}4 \times 10^6$	60
$2{\cdot}0 \times 10^7$	$1{\cdot}2 \times 10^7$	60
$2{\cdot}0 \times 10^7$	$8{\cdot}0 \times 10^2$	80
$2{\cdot}0 \times 10^7$	$9{\cdot}0 \times 10^1$	80
$2{\cdot}0 \times 10^7$	$9{\cdot}0 \times 10^1$	100
$2{\cdot}5 \times 10^7$	$2{\cdot}0 \times 10^6$	60
$2{\cdot}5 \times 10^7$	$4{\cdot}2 \times 10^6$	60
$2{\cdot}5 \times 10^7$	$9{\cdot}0 \times 10^1$	80
$2{\cdot}5 \times 10^7$	$9{\cdot}0 \times 10^1$	80
$2{\cdot}5 \times 10^7$	$9{\cdot}0 \times 10^1$	100
$3{\cdot}2 \times 10^7$	$1{\cdot}0 \times 10^7$	60
$3{\cdot}2 \times 10^7$	$2{\cdot}9 \times 10^7$	60
$3{\cdot}2 \times 10^7$	$2{\cdot}9 \times 10^5$	80
$3{\cdot}2 \times 10^7$	$8{\cdot}8 \times 10^5$	80
$3{\cdot}2 \times 10^7$	$9{\cdot}0 \times 10^1$	100
$3{\cdot}5 \times 10^7$	$3{\cdot}0 \times 10^7$	60
$3{\cdot}5 \times 10^7$	$2{\cdot}0 \times 10^7$	60
$3{\cdot}5 \times 10^7$	$1{\cdot}5 \times 10^3$	80
$3{\cdot}5 \times 10^7$	$4{\cdot}9 \times 10^5$	80
$3{\cdot}5 \times 10^7$	$9{\cdot}0 \times 10^1$	100

free-flowing powdered semi-products and viscous end products. Starch and cocoa powder, coating materials and walnut suspensions were used as model materials. The pastes studied were heated across their entire cross-section and heating periods much shorter than the conventional ones proved sufficient. At a frequency of 10^9 Hz a treatment of only 2–3 min proved adequate. Sterilisation was often combined with drying. The entire spectrum of bacteria was tested but neither *Escherichia coli* nor mould could be found. Sterilisation effects were noted at a temperature as low as 55–60°C but the most effective treatment conditions were found to be: temperature 85–95°C, and time 60–70 s.

Toshi and Muranaka[68] studied the effects of microwave heating of Castella pastries upon their bacterial activity and storage characteristics. Heating for 2 min to 170°C decreased an initial bacterium count of

TABLE 11

CHANGES OF THE PROPERTIES OF DRIED PASTE PRODUCTS DURING DIELECTRIC FINAL TREATMENT AT DIFFERENT TEMPERATURE LEVELS

	Final temperature of high-frequency treatment	Boiling time	Water absorption capacity	Measure of boiling to rags (%)	Comparison of treated and untreated samples with test, at 95% probability water absorption capacity	Boiling to rags
Untreated 'Shell'	—	19	135 ± 2·8	8 ± 2·3		
Treated 'Shell'	60	19	143 ± 1·4	8 ± 2·3	—	—
	80	19	146 ± 2·2	18 ± 2·2	—	—
	100	19	141 ± 1·4	18 ± 2·3	—	—
Untreated 'Square-Flake'	—	16	157 ± 1·5	0		
Treated 'Square-Flake'	60	16	159 ± 2·7	0	—	—
	30	16	164 ± 2·4	27 ± 3·5	—	+
Untreated 'Grains'	—	10	259 ± 2·8	33 ± 3·0		
Treated 'Grains'	100	10	265 ± 6·4	35 ± 5·0	—	—
Untreated 'Grains'	—	10	270 ± 1·4	5 ± 1·4		
Treated 'Grains'	100	10	278 ± 7·0	5 ± 2·0	—	—

+ Significant difference; — no significant difference.

$1\cdot97 \times 10^6$ to 10–47, recorded after a storage of 20 days at a temperature of 28°C. Similar treatment increased the storage life of conventional Japanese and Western pastries by a factor of 2–3. A brief microwave heating to 50°C increased the stability of Kamaboko and other fish-containing products. The thawing of frozen foodstuffs and the plasticisation of dry foodstuffs is also discussed briefly.

Pedenko et al.[69] studied the effects of continuous microwave treatment upon micro-organisms in various model materials. A Type G-3-10A generator, operated at 2375 MHz, was used. Considerable sterilisation was recorded. The lethal effect was twice as high as by conventional sterilisation. At higher temperatures sterilisation was much more powerful than that achieved by conventional means at comparable temperatures. Microbes were killed at 46–56°C in 8–20 min, while the control method required at least 60 min.

Chipley[70] discussed briefly the technique of microwave heating, then dealt with the effects upon micro-organisms of microwave radiation. Bacillus stearothermophilus and Aspergillus niger strains were used. There was no difference between the effects of microwave and conventional heating in the absence of water. Other experiments also proved that bacteria were killed by heat and that microwave treatment per se below the lethal temperature had no sterilising effects. Rogov et al.[71] studied the effects of microwave field strength upon the micro-organisms present in meat samples. They found that the microbial activity was reduced with increasing microwave energy input, and an increase in the rate of heating from 0·5°C/s to 0·6°C/s caused a significant increase in the elimination rate of bacteria. Kotula[72] discussed the literature dealing with the survival of contagious larvae of Trichinella spiralis in microwave-roasted pork. The time and temperature values required for the efficient elimination of T. spiralis are given, along with the details of USDA regulations. Excessive energy input leading to unequal heating represents a source of danger. The roles of the size of the slice, fat content and bones were discussed. He pointed out the importance of temperature equalisation following the process of microwave roasting and detailed other practical aspects, too.

Rosenberg and Bögl[73] mentioned the existence of several literature references indicating that microwave treatment was not as effective as the conventional one. Complete inactivation could be achieved only in the case of wet products. Only continuously operated, industrial-scale microwave systems ensure effective sterilisation and pasteurisation.

Domestic microwave ovens often stimulate the growth of micro-organisms and show increased bacterium counts. The paper emphasises the importance of chemical composition which can lead to varying dielectric constant and a protection for the micro-organisms. In their opinion the results can be improved by the alteration of the capacities and frequencies of the systems. Ishitani et al.[74] studied the effects of microwave heating upon the spores of fungi. There are several fungi which infect foodstuffs such as sponge cakes. They studied the behaviour of A. niger, Aspergillus restriturs, Paecilomyces varioti and Cladosporium herbarium strains. They concluded that compared with conventional heating, microwave heating has no specific effect. They also showed that in a dry glass bed microwave heating is ineffective against fungi.

Fruin and Guthertz[75] studied the survival of micro-organisms in microwave ovens, conventional ovens and slow heating systems. Meat slices were infected with E. coli, Clostridium perfringens, Streptococcus faecalis and S. aureus strains and various roasting techniques were tested. Part of the infected sample was used for bacteriological evaluation, while the other part was divided into three portions and roasted by microwave, conventionally and by a slow rise of temperature. During heating the average, minimum and maximum temperatures of the sample were determined. Immediately after roasting the samples were tested for overall and strain-specific bacterium counts. The survival rates as a function of temperature were calculated for each roasting method. They found that heat fluctuations were the highest for the microwave method and the smallest for the slowly heated sample. At a significance level of 0·05 there was no strain-specific difference in the survival rate obtained with the various techniques.

Dahl et al.[76] studied the effects of microwave reheating upon the micro-organisms in prepared and refrigerator-stored veal steaks and French beans. Both the veal steaks and the French beans were prepared in triplicate, according to the requirements of the 'Hazard Analysis by Critically Controlled Procedures' model. Veal steaks were heated to a final temperature of 66°C, then were stored at 6°C for 24 h and sliced. Beans were divided into portions following a storage of 24 h at 6°C. Each portion was separately heated in a microwave oven for 20, 50, 80 and 110 s. They found that when the heating time exceeded the 50 s value the number of micro-organisms effected was proportional to the duration of microwave treatment. Cunningham and Francis[77] studied the microwave-treatment-induced increases in the storage life of cooled chicken. The changes in the microflora of fresh chicken, especially the changes in the psychrotroph level, were examined after various forms of microwave treatment. They found that microwave treatments increased the storage life. Intense treatment for 20–40 s proved very effective for the elimination of psychrotrophic bacteria, such as *Pseudomonas* spp. coliforms, *Moraxella acinetobacter, Flavobacterium deborans, Alcaligenes faecalis, etc.* Thus, cooled chicken could be stored longer.

Holley and Timbers[78] examined the roles of microwave treatment and moisture upon the elimination of Nematospora infection of mustard seeds. They found that the fungal and Nematospora infection of mustard seeds can be eliminated at 87°C when the moisture content is 6·8%, and at 71°C when the moisture content is 9·7%. The moisture content also influences the activity of myronisase. They found that inactivation of myronisase required at least 10°C higher temperatures than the inactivation of yeast. They observed that at a moisture content of 10%, nematospores retained their vitality in untreated seeds for 2–8 years while storage was effected at 24°C.

Enami and Ikeda[79] took out a patent for the sterilisation of liquid foodstuffs, jam and ground meat, in which food is transported at elevated pressure through a conduit of low loss-coefficient where the foodstuff is heated by a microwave field. Sterilised food is transported to a packaging machine. The Mitsubishi Monsanto Chemical Company obtained a patent[80] for the microwave sterilisation of solid foodstuffs, fruits, vegetables, meat or fish. Solid food is mixed with liquids, such as water, at high temperature and is sterilised in closed containers in a microwave field. Stenström obtained a patent[81] for the heat treatment of heat-sensitive materials by electromagnetic radiation. Separate doses of the heat-sensitive materials are pre-treated to 50°C then their surface is cooled to avoid temperatures higher than 50°C. Then the product is pasteurised or sterilised by a high-frequency electromagnetic field.

4.3. Pre-cooking, Cooking and Baking

Cross and Fung[82] assert that microwave heating is fast and offers considerable savings in energy. They discuss the art and tricks of microwave cooking and the effects of microwave treatment upon the nutritional value of foodstuffs, Most publications maintain that microwave heating is accompanied by a higher loss of moisture than most conventional methods. Microwave heating affects proteins, lipids and minerals to a small extent. No reliable data are available relating to the behaviour of carbohydrates. There is a considerable body of information on the effects of microwave treatment of vitamins. It seems that there are only slight differences between the vitamin retention of microwave and conventional methods of cooking. In conclusion the paper states that there is no significant difference—from a nutritional point of view—between conventional and microwave cooking.

Glasscock et al.[83] studied the dielectric pre-cooking of vegetables prior to their storage in a frozen form. Both objective and sensory tests of vegetables blanched by microwave and conventional techniques were carried out. The tests, including broccoli, carrots, cauliflower, French beans and courgettes showed that there were no quality differences between the products obtained by the two different methods. Drake et al.[84] studied the effects of blanching and freezing methods upon the quality of certain vegetable products. Asparagus, French beans, green peas and sweet corn were blanched in boiling water, by steam and by microwave. Blanching by microwaves proved less favourable with regard to vitamin C retention, drip losses and colour. Kizevetter et al.[85] discussed the high-frequency cooking of Cucumaria japonica (sea cucumber) and published their rationalised cooking method. They concluded that high-frequency cooking required as short a time as 1/9 to 1/17·5 of the conventional method. The weight loss recorded in microwave cooking was below 40%, again 1/1·1 to 1/1·5 of what was observed in conventional cooking. Water retention was higher and 10–15% more of the biologically active glucosides could be preserved (0·023 and 0·028 instead of 0·020). Lipids were neither hydrolysed nor oxidised during high-frequency cooking of sea cucumber.

Dehne et al.[86,87] studied the effects of high-frequency cooking upon the nutritional values of foodstuffs of animal origin. They discussed the effects of microwaves upon the structural changes of foodstuff proteins, their solubility, the SH functional group content, pH, amino acid composition, water retention and digestibility. In the second paper they reviewed the effects of microwave heating upon the fat constituents of meat, neutral fats, phosphatides, cholesteroids, retinols, water losses, dry weight losses, minerals and trace elements. They also discussed the effects of microwave heating in the case of vitamins (thiamine, riboflavine, niacin, pyridoxine, retinol), additives and pesticide residues, nitrosamines, and the changes of the sensory characteristics of foodstuffs. Their results showed that microwave heating was not worse than conventional heating as regards the nutritional values of foods. Apparent differences could be often traced back to differences in methodology. The paper proposed further work for the comparison of conventional and microwave cooking.

Demeczky et al.[88] stated that cooking and smoking are the two most time-consuming steps of sausage production. They studied the use of continuous dielectric cooking to replace the batchwise, conventional methods of cooking. The paste used to produce sausages was fairly homogeneous and had a high moisture content. They determined the ε_r' and ε_r'' values, between 10 and 90°C temperatures, of the paste used to produce 'Parisian sausage', the most popular sausage in Hungary. They found that ε_r' hardly changed at all. The value of tan δ slightly increased up to 60°C, then it decreased again. The paste was continuously forced through a 100 mm internal diameter polypropylene tube. Twin spiral electrodes, located at the circumference of the tube, were used to achieve a homogeneous heating regime for the entire mass. Energy input and the material flow were controlled to maintain a final temperature of 70–72°C. The energy requirement was 0·21 kWh/kg sausage. The phosphatase test was used to compare the degree of cooking of the conventionally prepared and microwave cooked sausage. The total micro-organism count was also determined along with detailed sensory evaluations. It could be concluded that there was no significant difference between the products obtained by the two different methods. Based on the results obtained with the laboratory-scale high-frequency cooking unit (Fig. 7) a plant-scale unit has been designed (Fig. 8). Detailed cost analysis showed that the high-frequency cooking units paid for themselves in 0·1–0·2 years.

Cremer[89] compared the energy consumption and sensory values of scrambled eggs and beef paté heated by conventional and microwave technologies. Using 96 samples each they found that the conventional methods required significantly more energy ($P < 0·01$) for heating than the microwave technology. Scrambled eggs required 30·25 MJ and 9·9 MJ energy while beef patés needed 35·15 MJ and 9·88 MJ, respectively. Sensory evaluation showed that microwave heating was significantly better for scrambled eggs ($P < 0·01$), but conventional heating proved better for paté.

Andres[90] reported on the use of aromas designed to make the taste

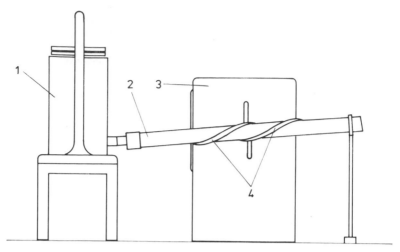

FIG. 7. Schematic of a continuously operated, laboratory-scale, high-frequency Parisian sausage-cooking unit. (1) Continuously operated sausage filling unit; (2) cooking tube; (3) high-frequency generator and (4) spiral electrodes (around the polypropylene cooking tube).

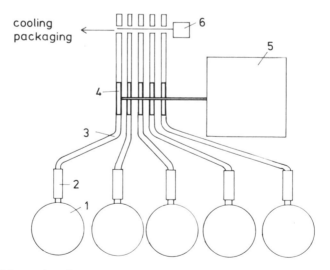

FIG. 8. Schematic of a continuously operated, plant-scale high-frequency Parisian sausage-cooking unit. (1) Storage tank (agitated, evacuated); (2) feeding pump; (3) cooking tube (polypropylene); (4) electrode; (5) high-frequency generators and (6) cutting system.

of microwave roasted beef similar to that roasted at higher temperatures. The process is quite advantageous because Maillard reactions do not occur. Riffero and Holmes[91] pressed pre-rigor beef samples into forms and heated them both conventionally and by microwaves. They found that, compared with conventionally prepared samples, pre-rigor pressed samples had a higher moisture content, higher pH, better colour and texture ($P < 0.05$). The moisture content, drip-loss, inner red value were decidedly better for the dielectrically heated samples than for the conventionally baked ones. There was no difference either in the smell or the juice content of the various samples ($P < 0.05$). Ray et al.[92] examined the effects of various electrical stimulation methods and pre-treatment methods (freezing, thawing, pre-conditioning) upon the taste, chewing and slicing characteristics of microwave heated pre-rigor meats. Nader et al.[93] studied the mutagenic factors for hot-plate roasted and microwave roasted beef. Beefsteaks were prepared in parallel on a hot plate and in a microwave oven (2450 MHz). The mutagenic activity of the surface samples of both beefsteaks was determined using Salmonella typhimurium ZA 98 TA 100 strains, with and without S-9 microsomal pre-treatment. With S-9 activation, beefsteaks produced on hot plates showed definite mutagenicity which increased with the duration of roasting. No similar effects were noted for microwave roasted beefsteaks either at normal roasting times or during three times longer periods.

The advantages of microwave heating compared to conventional heating have been reviewed in ref. 94. Applications for hamburgers, sausages, pre-weighed meat products, canned meat products, and for the pre-production processing of frozen meat were discussed. The average capacity of such systems was 2700 kg/h. At present, some 10^6 kg products are produced, all over the world, using this technology. McNeil and Penfield[95] studied various baking possibilities and their effects upon the quality of roasted turkey. Energy utilisation was much better in the microwave and convection-type ovens than in the conventional ovens. There was no difference between the various roasting methods regarding the roasting losses and moisture contents. Neither did the degree of roasting, the appearance, aroma, juice content and flavour of the turkeys differ after storage in a refrigerator for 24 h. They found that microwave roasted turkeys were not quite as soft as the conventionally prepared ones. Consumers (72 consumers and several repetitions) found no difference between the samples prepared by the different methods.

Nekrutman et al.[96] studied the combined application of microwave and infra red heating in the case of meat packaged in plastic films. They found that, compared with the conventional roasting method, the combined method increased the storage life of roasted meat at 4–8°C by 2 months. The combined infra red–microwave method proved much more suitable for pasteurisation than the conventional methods. Heating by the combined method to 85–90°C proved much more effective than even higher temperatures achieved by the conventional methods. Heath and Reilly[97] studied the uptake of plasticisers during heating of foodstuffs packaged in plastic films. Using models, they studied the transfer of tributyl acetyl citrate into poultry meat during microwave heating. Based on the results of infra red spectrophotometric studies they concluded that migration was significant after a cooking period of 8 min, and that the lipid fraction was responsible for both the retention and migration of tributyl acetyl citrate. When the lipid concentration of the model material was increased the amount of tributyl acetyl citrate retained became higher.

Cross and Fung[98] studied the effects of dielectric cooking upon six major food constituents comparing them to the effects of conventional cooking. Proteins, lipids, water, carbohydrates, minerals and vitamins were studied. They concluded that the two different heating methods showed identical effects in each group studied. Baldwin[99] discussed microwave ovens and the effects of microwave heating upon the most important constituents of foodstuffs. He also dealt with the microbiological aspects of microwave heating, the debate on Trichina, the energy conservation aspects and the federal legislations relating to microwave ovens. He also discussed the effects of microwave fields upon the operation of heart pacemakers.

Demeczky et al.[100], Runtág and Demeczky[101] and Demeczky and Márkus[102] studied the conventional, combined infra red and microwave methods of continuous bread baking. They also developed experimental systems and tested them at the Central Food Research Institute. They determined the water uptake of various flours, the electrophysical characteristics of bread doughs. Measurements were carried out in the 30–98°C range, which is of prime importance in baking. ε_r' and ε_r'' values increased up to 70°C, then a slight decrease followed. The temperature and weight loss of bread dough as a function of the power input and heating time were studied. Based on these data they developed a continuously operated plant-scale oven. Bread baked in this oven was passed through either an infra red tunnel oven or a 325°C conventional oven where the crust was formed. Based on the bulk weight, structure and sensory evaluation of bread they recommend that 4–5 min of high-frequency baking be combined with 6–7 min in a conventional oven. The average specific energy consumption of high-frequency baking is 0·16–0·18 kWh/kg product, that of the post-treatment in the conventional oven is 0·10–0·12 kWh/kg product. Lower grade flours could be also used to produce excellent bread in the combined oven. For example, flours with high amylase activity which proved unsuitable for bread-making in conventional ovens gave good-quality bread in the combined oven. The specific volume of bread produced by the combined technology is some 20–30% higher than that of the control bread loaves produced by the conventional methods (Fig. 9). The direct baking costs of the combined technology are only 2/3 of the conventional technology. Using this combination technology the production capacity of existing baking lines can be considerably increased with a minimum capital investment and space uptake. Enkina et al.[103,104] studied the effects of high-frequency heating upon the physical characteristics of bread dough and its baking characteristics. Experiments were carried out with a first class flour of 11·3% moisture content and 3·8 acid value. The gluten content of flour was 30·6%. Its elongation upon pulling was 20 cm. Heating of the flour for 90 s to 333 K did not change its gluten content, but further heating to 349 K decreased it by 2·4%. Heating for 60 s reduced the elongation by 3 cm. Further heating caused further decreases in the gluten concentration. The amount of carbon dioxide liberated during heating increased to 2–8·7%. They concluded that the baking characteristics of

FIG. 9. Cross-section of breads. Left: bread baked by the traditional method; right: bread baked by the high-frequency method.

flours can be controlled in high-frequency fields as required by the technology. In another series of experiments flour was heated to 35, 40, 60 and 70°C in a microwave oven, cooled to ambient temperature and the dough was produced. The standard microwave oven technique was used to bake the bread loaves. Flours pre-treated to 60°C gave better bread than the control sample. Even a slight pre-treating for 6 s to 40°C improved the quality of both the flour and the bread made thereof.

McArthur and D'Appolonia[105] studied the effects of microwave radiation upon the baking characteristics of flour obtained from red spring wheat. Experimental flour samples were treated by microwaves and stored for 6 months at room temperatures. Treatments were carried out with 625 W energy for 480 s. They found that treatments longer than 280 s had deleterious effects on the quality of bread prepared from the treated flour. High-power microwave irradiation yielded anomalous farinograms with twin peaks. Moderate microwave treatments had no deleterious effects, in fact both the bulk weight and other quality indicators were improved. Selyagin and Stephanovits[106] studied the kinetics of microwave baking of biscuit. They found that high-frequency baking is 5–10 times faster than conventional baking. In layer units the temperature profiles are more uniform and the moisture loss is lower. In high-frequency fields heating proceeds from the inside and evaporation takes place at the surface. Pei[107] examined the characteristics of microwave baking. He found that the best results were obtained with the combination of microwave and conventional techniques.

Hulls[108] recalled that radio frequency heating systems were introduced into the United Kingdom some 40 years ago and they had long proved their worth. At present, some 90% of all dielectric systems operate in the radio frequency range, predominantly at 13·56 MHz. Microwave systems are the results of newer developments and, as such, are only at the beginning of their career. In his opinion, the present situation of the world economy demands a careful choice of the heating technology. Apart from the description of the general application fields of radio frequency dielectric heating systems, he singles out industrial systems used for the post-baking of crackers and cookies. These systems ensure a uniform moisture content, within ±0·2%, without rendering the products fragile. Radio frequency post-baking of crackers and baker's ware is a well established industry. More than 100 units of 40–100 kW capacity are in use in the United

States and Europe. He also singles out the significance of microwave units used to maintain the temperature of frozen meat blocks at 0°C. Such systems are well-spread in the United States, but an experimental, batchwise 30 kW unit and a 60 kW continuously operated tunnel oven have also been built in the United Kingdom.

Albrecht and Baldwin[109] studied the changes of aroma content, thiamine and riboflavin concentration and lipid oxidation during reheating of roasted frozen meat by conventional and microwave methods. 2·7 kg meat was stored at -23°C, placed in to a conventional electric oven (163 ± 2°C) and heated to an internal temperature of 75 ± 2°C. The meat sample was removed from the oven and sliced immediately to slices of 0·6 cm thickness. These slices were used for the reheating experiments, chemical analysis and sensory evaluations. Reheating was also realised in two microwave ovens of 525 W and 1160 W capacity, respectively (both operated at 2450 MHz), and in a conventional electric oven (163 ± 2°C). Sensory evaluation tests and chemical analysis results were subjected to statistical analysis. They found that both the conventional and the microwave ovens gave identical products in the reheating step. Neither the sensory evaluation nor the vitamin retention values were better for the microwave ovens which required much shorter heating periods. Lipid oxidation and flavour and aroma changes occurred during storage of meat. Lipid oxidation was detected in the sensory evaluations as warmed-over-aroma (WOA) and warmed-over-flavour (WOF). The time and energy requirements of reheating were more favourable in the case of microwave ovens.

Miller et al.[110] carried out comparative studies on the feather release characteristics of poultry. A large number of experiments were carried out both with hot water and microwaves. They found that the microwave technique was suitable for feather release in poultry and the danger represented by the infected boiling water could be avoided. One of the existing problems of the microwave technique is the reduction of the internal heating of the body. Some 61·2 kJ energy can be conserved per chicken.

Intensive research during the past years for the application of dielectric and microwave heating resulted in a number of patents. Often, dielectric heating is combined with infra red irradiation or resistance heating. Several patents were issued for such automated systems or programmed systems. Due to the space limitations of this chapter only a few examples can be mentioned at random. Schiffmann *et al.*[111] obtained a patent for combined bread baking on metal pans. The combination of conventional and microwave baking resulted in decreased baking times. The baking losses were also decreased. Using microwave generators of 915 MHz and 2400 MHz very even baking could be achieved. Meinass[112] took out a patent for the combination of conventional and microwave ovens in order to reduce both the energy and time requirements of cooking and baking. Larsen and Duncan[113] obtained a patent for an air circulating system applied in a combined conventional and microwave oven. Tanaka *et al.*[114] obtained a patent for a cooking system in which food is heated either by resistance heating, microwave heating or a combination of the two. Food is placed into a revolving container ensuring even heating. The simultaneous application of resistance heating and microwave heating offers new possibilities for food preparation. A patent was issued to Maxton,[115] describing a system in which meat, e.g. chicken pieces, is moved through a container of cooking oil and subjected to microwave radiation. The use of the microwave technology ensures fast frying, yet, at least half of the energy used can be secured from cheaper sources, e.g. gas. Bowen *et al.*[116] described a microwave coffee pot in which a small microwave oven is used to produce steam.

Noda's patent[117] describes an automatic mechanism for the control of energy input to the material. Horinouchi[118] described a microprocessor-based control system for microwave ovens. Takasaka *et al.*[119] developed a magnetic card system for the control of the cooking regime. Buck[120] described a method and apparatus for the control of the cooking time using telemetric sensing of the temperature of the foodstuff and microprocessor-based control circuitry. The patent of Doi and Kashiwagi[121] is based on a digital control circuit, combined with a microwave oven and a digital display, for the control of the microwave generator and the indication of cooking conditions and/or occasional errors.

The high-frequency heating system described in Noda's patent[122] has a program card in which the energy levels corresponding to the required cooking temperatures and times are recorded. Data contained in the card are read out and stored in a memory. Heating is commenced according to these data stored in the memory and either the cooking time or the cooking temperature can be displayed.

A disposable unit is described in the patent issued to Kubialovicz[123] to be used with foodstuffs which exude juice, for example ham. It is constructed of non-metallic materials and prevents the juice formed during cooking from reaching the surface of the microwave oven.

Andres[124] reports that glass factories studied the suitability of glass containers for the production of microwave-sterilised foodstuffs. Experiments showed that the normal glasses are transparent to the microwave radiation. Labels can be left on the bottles and formed polystyrene labels make the handling of hot bottles easier. The glass containers do not interfere with the distribution of the microwave field. The containers are reusable.

4.4. Defrosting of Frozen Products

In 1963 Bengtsson[125] carried out fundamental comparative laboratory experiments for the defrosting of frozen beef and cod fish using a continuously operated dielectric defroster. Experiments at 35 MHz were carried out between plane electrodes while those at 2450 MHz were carried out with a parabolic oven. Regular frozen blocks were used in both experiments. After defrosting the average temperature of beef blocks was 0·4°C (at 35 MHz) and 0·2°C (at 2450 MHz) while the values for the cod fish blocks were −0·7°C and −0·6°C, respectively. Drip-loss for the frozen beef blocks was 0·3 and 0·5%, for the cod fish blocks 0·7 and 0·6%, respectively. Samples tested organoleptically after cooking did not differ depending on the defrosting method used. Both frequencies proved adequate for defrosting but the higher power density at 2450 MHz led to shorter defrosting periods. However, due to the limited depth of penetration at this frequency, the thickness of sample had to be kept below 4 cm. A plant-scale experimental system has been developed in the Hungarian Central Food Research Institute[126] for the continuous defrosting of frozen liver and deboned frozen meat blocks. In this work the conditions, energy consumption, defrosting period and weight loss of conventional defrosting techniques were surveyed. High-frequency defrosting experiments were carried out with a 13·65 MHz generator supplying 20 kW of useful power. Both the laboratory and plant-scale experiments showed that high-frequency defrosting offers a saving of 1·5–5% in weight-loss, due to reduced drip-loss. The consistency of the defrosted product could be controlled via its final temperature as required by the successive processing steps. Plant-scale experiments showed that high-frequency defrosting offered flexible operation and proved more suitable than the conventional techniques both in hygienic and economic aspects. The schematic of the plant-scale unit is shown in Fig. 10. The results of the tests are tabulated in Table 12.

A method has been developed in the Hungarian Central Food Research Institute[127] for the rapid, controlled defrosting of frozen

FIG. 10. Experimental meat-defrosting unit.

fruits, vegetables and purees. The method ensures that there is no defrosting-related weight-loss. With this technique double sterilisation could be avoided. This is often a necessary pre-condition for the maintenance of product quality and is required by export considerations. A continuously operated, laboratory-scale high-frequency defrosting system has also been developed for the defrosting of egg mixture, frozen in polyethylene containers.[128] The ε_r' and ε_r'' values of egg mixture as a function of temperature were determined in the $-10°C–30°C$ range. A continuously operated plant-scale system has also been built for the defrosting of polyethylene cans containing 5·4 kg of frozen egg mixture. The specific energy consumption of heating from $-10°C$ to $10°C$ was also determined. Experiments showed that, compared with the conventional method, a weight-loss of 3% could be avoided and the product obtained by high-frequency defrosting proved superior from a hygienic point of view, too.

TABLE 12

DRIP-LOSSES OF FROZEN LIVER BLOCKS DURING CONVENTIONAL AND HIGH-FREQUENCY DEFROSTING AND HEATING TO 10°C

Test series	Freezing and storage loss calculated to pre-freezing weight basis	Drip-loss in wt/wt related to pre-freezing weight			
		High-frequency defrosting	Refractory content of drip	Conventional defrosting	Refractory content of drip
I	0·14 ± 0·02	1·28		3·59	
II	0·12 ± 0·02	1·59		3·91	
III	0·18 ± 0·03	1·32		3·86	
IV	1·38 ± 0·06	3·22	18·1	5·60	20·7
V	1·53 ± 0·12	4·26	17·9	6·54	19·3
VI	1·69 ± 0·10	4·62	17·1	7·47	19·6

Economic calculations showed that high-frequency defrosting costs were half of the costs of conventional defrosting methods. The investment costs of a 2 t d^{-1} capacity defrosting unit can be recovered in 1·2 years. An economic laboratory-scale method has been developed in the Hungarian Central Food Research Institute for the defrosting and partial heating of frozen fish blocks, too. Frozen fish blocks, treated by high-frequency radiation, could be separated in the −1·5 to −1·0°C temperature range. Thus, frozen fish could be repackaged without the danger of spoiling and dispatched, through a chain of suitable coolers, to retailers. The energy consumption of heating was comparatively low, 110 kWh/t.

Demeczky et al.[129] studied the heating by high-frequency radiation to 5°C of salami filled at −5°C. The enthalpy changes associated with the heating of salami from −5°C to 5°C were also determined. They found that, on the average, 48·6–61·5 kJ energy was required for each kilogram of salami. The conductance of various salamis at 13·56 and 30 MHz as a function of temperature was also determined. They found that the resistance dropped significantly with increasing frequency and temperature. When the energy input level was maintained in such a manner that heating took 15 min, the standard deviation of product temperature was less than ±1°C.

Karabodzhov[130] studied the microwave heating of frozen ready-made foods. The defrosting and heating time and the weight-loss of kebab and veal steaks with green bean garnish were determined. Samples of 10, 20, 30, 40 and 50 mm thickness with 380–2260 g weight were prepared. Samples were kept at −18°C for 48 h. Defrosting and heating time correlated directly with the thickness, weight and electrophysical characteristics of the sample. Vavrik,[131] studied the microwave defrosting of 25·4 kg frozen butter blocks. Butter blocks of −18 to −14°C were heated in 18 min to 6°C. On average, 2·22 kW/h energy was required for the heating of frozen butter blocks. Ohlsson[132] discussed the heat conductance phenomena in frozen foodstuffs during microwave defrosting, with special emphasis on the extremely low heat conductance of the frozen product in comparison with the heat conductance of the fresh product. These low heat conductance values limit the widespread application of microwave heating for defrosting purposes. Experiments carried out with frozen meat samples proved that defrosting taking several hours to complete leads to the best product quality. Compared with this, microwave techniques seem very favourable both in product quality and processing time.

Oterbajn[133] states that the use of frozen meat in meat products is hindered by the fact that defrosting takes twice as long as freezing and that homogeneous defrosting is difficult to ensure. Conventional defrosting often leads to temperature gradients between the interior and exterior of the product as high as 10°C, and the drip-loss is also significant. Microwave heating of meat blocks kept at −20° to −18°C is advantageous, because it is rapid, the temperature is even throughout the entire block and there is no drip-loss. The system called 'Raytherm' is also described in this paper. Its nominal capacity is 1000–3000 kg h^{-1} using frozen meat blocks of 25–30 kg. The advantages of the system include continuous operation, even temperature distribution across the entire block, small space requirement, ease of cleaning, and the easy removal of cardboard packaging. The danger of spoiling during defrosting is also eliminated and meat quality is better because of the lack of drip-loss.

5. A BRIEF SUMMARY OF THE APPLICATION OF HIGH-FREQUENCY AND VERY-HIGH-FREQUENCY DIELECTRIC HEATING IN FOOD PROCESSING

Conventional fuel sources are likely to be depleted soon. The development of new sources of energy, the introduction of solar and nuclear energy stimulate the more efficient utilisation of electric power. In this respect dielectric techniques might play an important role.

High-frequency and very-high-frequency dielectric techniques have found increasing application in the food processing industries during the past two decades following the results of theoretical and practical studies. For food processing it is very important that the selection of the suitable conditions (geometry, size, frequency, field strength, duration of treatment) of dielectric heating allows for very rapid, homogeneous, economic heating of various foodstuffs. Dielectric techniques are also suited for automated, continuous operation. The elimination, or at least reduction, of local temperature gradients allows for considerable raw material conservation and improved quality via the exclusion of harmful, burnt products. Rapid energy input decreases the size of production facilities and offers improved hygienic conditions.

It can be concluded from the results of dielectric heating research that the rational selection of the conditions of dielectric energy input and its occasional combination with conventional technologies lead to products which meet the quality requirements of existing products and open the field for new products.

Due to these advantages, investment costs of dielectric installations can, in general, be recovered in 0·2 to 2 years.

REFERENCES

1. BENGSSTON, N. E., MELIN, J., REMI, K. and SÖDERLIND, S., *J. Sci. Food Agri.*, 1963, **8**, 592–604.
2. SHARMA, M. N., *J. Sci. Ind. Res. Sect. B.*, 1960, **19**, 5–7.
3. JASON, A. C. and SANDERS, H. R., *Food Technol.*, 1962, **16**, 101.
4. OHLSSON, T. H., BENGTSSON, N. E. and RISMAN, P. O., *J. Microwave Power*, 1974, **9**, 129–45.
5. NELSON, S. O., SODERHOLM, L. H. and YUNG, F. D., *Agri. Eng.*, 1953, **34**, 608–10.
6. NELSON, S. O., *Trans. ASAE*, 1965, **8**, 38–47.
7. JORGENSEN, J. L., EDISON, A. R., NELSON, S. O. and STETSON, L. E., *Trans. ASAE*, 1970, **13**, 18–20, 24.
8. STETSON, L. E. and NELSON, S. O., *Trans. ASAE*, 1970, **4**, 491–5.
9. CORCORAN, P. T., NELSON, S. O., STETSON, L. E. and SCHLAPHOFF, C. W., *Trans. ASAE*, 1970, **13**, 348–51.
10. NELSON, S. O., *Trans. ASAE*, 1973, **16**, 384–400.
11. NELSON, S. O., SCHLAPHOFF, C. W. and STETSON, L. E., *J. Microwave Power*, 1973, **8**, 13–22.
12. NELSON, S. O., *Trans. ASAE*, 1973, **16**, 902–5.
13. TINGA, W. R. and NELSON, S. O., *J. Microwave Power*, 1973, **8**, 23–65.
14. NELSON, S. O., SCHLAPHOFF, C. W. and STETSON, L. E., *IEEE Trans. Microwave Theory Tech.*, *MTT*, 1974, **22**, 342–3.
15. NELSON, S. O., STETSON, L. E. and SCHLAPHOFF, C. W., *IEEE Trans. Instrum. Meas.*, 1974, **23**, 455–60.
16. NELSON, S. O., *Digest of the 1978 Microwave Power Symposium*, June 28–30, Ottawa, Ontario, Canada.
17. NELSON, S. O., *J. Microwave Power*, 1978, **13**, 213–8.
18. NELSON, S. O., *J. Food Process. Preserv.*, 1979, **2**, 137–54.
19. NELSON, S. O., *Trans. ASAE*, 1979, **22**, 950–4.
20. NELSON, S. O., *Trans. ASAE*, 1979, **22**, 1451–7.
21. NELSON, S. O., *Trans. ASAE*, 1981, **24**, 1573–6.
22. NELSON, S. O., *Trans. ASAE*, 1982, **25**, 1045–9.
23. NELSON, S. O., *J. Microwave Power*, 1983, **18**, 143–52.
24. MYNOV, V. A., KRAVCHENKO, V. M. and BELOUSOV, S. A., *Khlebopek. Konditer. Prom.*, 1981, **11**, 39–41.
25. PONOMAREVA, N. A., KORNENA, E. P., NIVOROZHKIM, L. E. and ARUTYUNYAN, N. S., *Izv. Vysh. Uchebn. Zavad. Pishch. Tekhnol.*, 1982, **4**, 106–8.

26. ROGOV, I. A., ILYUKHIN, V. V., BABAKIN, B. S. and LEBEDEV, N. M., *Myasn. Ind. S.S.S.R.*, 1982, **9**, 32–3.
27. RISMAN, P. O. and BENGTSSON, N. E., *J. Microwave Power*, 1971, **6**, 101.
28. BENGTSSON, N. E. and RISMAN, P. O., *J. Microwave Power*, 1971, **6**, 107.
29. OHLSSON, T. and RISMAN, P. O., *J. Microwave Power*, 1978, **13**, 303.
30. OHLSSON, T., *SIK*, 1975, **380**, 1–76.
31. SIMOVYAN, S. V. and POTAPOV, V. A., *Izv. Vyssh. Uchebn. Zaved. Pishch. Tekhnol.*, 1982, **6**, 61–5.
32. PFÜTZNER, H., FIALIK, E., KRAUSE, E., KLEIBEL, A., *Proc. Eur. Meeting of Meat Res. Workers*, 1981, **27**, **1**, **A14**, 50–3.
33. KLEIBEL, A., PFÜTZNER, H. and KRAUSE, E., *Fleischwirtschaft*, 1983, **63**, 322–3, 326, 328, 384.
34. DEMECZKY, M., ALMÁSI, E. and KÁRPÁTI, GY., *Method and apparatus for the recovery of non-denaturalized protein-fiber from fatty animal tissue*. Hungarian Patent 155423, 1965.
35. DEMECZKY, M., Hungarian Central Food Research Institute, Budapest, *Research Report*, 1968. Development of a continuously-operated high-frequency fat-frying line. pp. 1–78.
36. YAMAGUCHI, Z., *N. Food Ind.*, 1982, **24**, 1–5.
37. CADDEDA, S., *Proceedings from the second biennial Marschall International Cheese Conference*, Madison, Wisconsin, USA, 1981, September 15–18, 176–179.
38. ENTREMANT, J. and LEVARDAN, R., U.K. Patent Application 2074601 A, 1981.
39. POUR-EL, A., NELSON, S. O., PECK, E. E., TJHIO, B. and STETSON, L. L., *J. Food Sci.*, 1981, **46**, 880–5, 895.
40. DEMECZKY, M., GONDAR, J. and JÁKY, M., *Olaj, Szappan, Kozmetika*, 1960, **11**, 49–55.
41. DEMECZKY, M. and RUNTÁG, T., Hungarian Central Food Research Institute, Budapest, *Research Report*, 1970. The effects of high-frequency dielectric treatment upon the oil-extraction characteristics of seeds. pp. 1–69.
42. MAHESVARI, P. N., *Diss. Abstr. Int. B.*, 1981, **42**, 955.
43. PEDENKO, L. N., LERINA, I. V., BELITSKII, B. I., BELOVA, T. S. and TKACHENKO, L. F., *Ratsional'noe Pitanie*, Respublikanskii Mezhvedomstvenny Sbornik, 1982, No. 17, 102–4.
44. DEMECZKY, M. and GONDÁR, J., *Ber. Inst. Tabakforsch. Dresden*, 1961, **BVIII/I**, 102–13.
45. DEMECZKY, M., *Élelmiszertudomány*, 1968, **2**, 1–2, 80–2.
46. DEMECZKY, M., *UIE 9th Int. Congr.*, Cannes, 1980. Session No. 4. **111.A5**, 1–12.
47. DEMECZKY, M., *Élelmez. Ipar*, 1982, **36**, 441–9.
48. AKAHASHI, R. and MATASHIGE, E., *J. Japan. Soc. Food Sci. Technol.*, 1982, **29**, 587–92.
49. KAMAI, I., KIKUCHI, S., MATSUMOTO, S. and OBARA, T., *J. Agric. Sci., Tokyo Nogyo Daikagu Nogatu Shuho*, 1981, 90th Anniversary Edition, 150–68.

50. Leonhardt, G. F. and Gomez, A. M. F., *Cien. Technol. Aliment.*, 1981, **1**, 37–50.
51. Fregni, Z. E. and Leonhardt, G. F., *Cien. Technol. Aliment.*, 1982, **2**, 151–63.
52. Demeczky, M., Márkus, P. and Gross, J., Hungarian Central Food Research Institute, Budapest, *Research Report*, 1978. Dielectric final drying in vacuum for the production of low-moisture-content pasta-products. pp. 3–6.
53. Brighenti, F., Cavallini, A., Perlaska, M., Finazzi, M. and Cantoni, C., *Technol. Aliment.*, 1981, **4**, 7–15.
54. Meisel, N., *Microwave Power Symp.*, Monaco, 11 June 1979. *Microwave Power Appl.*, 93–6.
55. Ang, T. K., Pei, T. C. D. and Ford, J. D., *Chem. Eng. Sci.*, 1977, **32**, 1477–89.
56. Arsem, H., Shults, G. W. and Turmy, J. M., *Proc. Eur. Meet. Meat Res. Work.*, 1980, **26**, **1**, **E21**, 245–7.
57. Sunderland, J. E., *J. Food Process. Eng.*, 1980, **4**, 195–212.
58. Sunderland, J. E., *Food Technol.*, 1982, **36**, 50–2, 54–6.
59. Demeczky, M., *Élelmiszertudomány*, 1968, **2**, 85–8.
60. Kuznetsova, L. S., Kovaleva, L. S., Selyagin, V. I. and Konkina, G. A., *Khlebopek. Konditer. Prom.*, 1982, **8**, 33–4.
61. Krüger, E. and Groenick, E., *Monatschr. Brau.*, 1981, **34**, 173–8.
62. Luter, L., Wyslouzil, W. and Kashyap, S. C. *Can. Inst. Food Sci. Technol. J.*, 1982, **15**, 236–8.
63. Gerling, J. E. and O'Meara, J. P., USA Patent US 4326114, (1982).
64. Demeczky, M., *Proc. Int. Congr. Food Sci. Technol.*, 1974, **IV**, 11–20.
65. Bookwalter, G. N., Shukla, T. P. and Kwolek, W. F., *J. Food Sci.*, 1982, **47**, 1683–6.
66. Demeczky, M., Márkus, P., Hungarian Central Food Research Institute, Budapest, *Research Report*, 1979. Experiments to decrease the overall germ count of dried pasta-products. pp. 1–20.
67. Nosikov, V. F., Ostapenkova, T. N. and Dimitrieva, L. P., *Khlebopek. Konditer. Prom.*, 1981, **11**, 34–5.
68. Toshi, T. and Muranaka, T., *N. Food Ind.*, 1982, **24**, 12–7.
69. Pedenko, A. L., Belitskii, B. J., Lerina, I. V., Makeev, Yu. V. and Kutashev, V. N., *Izv. Vyssh. Ucsebn. Zaved. Pishch. Tekhnol.*, 1982, **5**, 54–6.
70. Chipley, J. R., *Adv. Appl. Microbiol.*, 1980, **26**, 129–45.
71. Rogov, I. A., Papkova, V. B., Biletova, N. V. and Nekristman, S. V., *Myasn. Ind. S.S.S.R.*, 1982, **4**, 35–6.
72. Kotula, A. W., *Proc. Annu. Reciprocal Meat Conf. Am. Meat Sci. Assoc.*, 1983, **35**, 77–80.
73. Rosenberg, U. and Bögl, W., *Fleischwirtschaft*, 1982, **62**, 1182–7.
74. Ishitani, T., Koyo, T. and Yanai, S., *Rep. Nat. Food Res. Inst.* (*Japan*), 1981, **38**, 102–5.
75. Fruin, J. T. and Guthertz, L. S., *J. Food Prot.*, 1982, **45**, 695–8, 702.
76. Dahl, C. A., Matthews, M. E. and Marth, E. H., *Eur. J. Appl. Microbiol. Biotechnol.*, 1981, **11**, 125–30.

77. CUNNINGHAM, F. E. and FRANCIS, C., *Feedstuffs*, 1982, **54**, 23–4.
78. HOLLEY, R. A. and TIMBERS, G. E., *Can. Inst. Food Sci. Technol. J.*, 1983, **16**, 68–75.
79. ENAMI, Y. and IKEDA, T., (Sanyo Electric Co. Ltd) UK Patent Application, 1982, 2098040 A.
80. MITSUBISHI MONSANTO CHEMICAL CO. UK Patent Application 1981. 2061085 A.
81. STENSTRÖM, L. A., (ALFA-LAVAL AB) United States Patent Reissue 1981, 30780.
82. CROSS, G. A. and FUNG, D. Y. C., *Crit. Rev. Food Sci. Nutr.*, 1982, **16**, 355–81.
83. GLASSCOCK, S. J., AXELSON, J. M., PALMER, J. K., PHILIPS, J. A. and TAPER, L. J., *Home. Econ. Res. J.*, 1982, **11**, 149–58.
84. DRAKE, S. R., SPAID, S. E. and THOMPSON, J. B., *J. Food Qual.*, 1981, **4**, 271–8.
85. KIZEVETTER, I. V., SAVATEEVA, L. YU. and NOVIKOVA, N. V., *Rybn. Khoz.*, 1981, **10**, 73–4.
86. DEHNE, L., BÖGL, W. and GROSSKLAUS, D., *Fleischwirtschaft*, 1983, **63**, 231–4, 236–7.
87. DEHNE, L., BÖGL, W. and GROSSKLAUS, D., *Fleischwirtschaft*, 1983, **63**, 1206–11.
88. DEMECZKY, M., RUNTÁG, T. and BALÁZS, D., Hungarian Central Food Research Institute, Budapest, *Research Report*, 1969. Continuous cooking of red-meat sausages by high-frequency energy input. pp. 1–65.
89. CREMER, M. L., *J. Food Sci.*, 1982, **47**, 871–4.
90. ANDRES, C., *Food Process.*, 1982, **43**, 70–1.
91. RIFFERO, L. M. and HOLMES, Z. A., *J. Food Sci.*, 1983, **48**, 346–50, 374.
92. RAY, E. E., BERRY, B. W., LOUKS, L. J., LEIGHTON, E. A. and GARDNER, B. J., *J. Food Prot.*, 1983, **46**, 954–6, 964.
93. NADER, C. J., SPENCER, L. K. and WELLER, R. A., *Cancer Letters*, 1981, **13**, 147–51.
94. ANON., *Food Eng. Int.*, 1982, **7**, 61, 63.
95. MCNEIL, M. and PENFIELD, M. P., *J. Food Sci.*, 1983, **48**, 853–5.
96. NEKRUTMAN, S. V., PAPKOVA, V. B. and DENISKINA, T. G., *Myasn. Ind. S.S.S.R.*, 1982, **11**, 22–3.
97. HEATH, J. L. and REILLY, M., *Poult. Sci.*, 1981, **60**, 2258–64.
98. CROSS, G. A. and FUNG, D. Y. C., *J. Environ. Health*, 1982, **44**, 188–93.
99. BALDWIN, R. E., *J. Food Prot.*, 1983, **46**, 266–9.
100. DEMECZKY, M., RUNTÁG, T. and BALÁZS, D., Hungarian Central Food Research Institute, Budapest, *Research Report*, 1971. Application of high-frequency electromagnetic radiation for the baking of bread and fine bakery products. pp. 1–74.
101. RUNTÁG, T. and DEMECZKY, M., *Acta Aliment. Acad. Sci. Hung.*, 1973, **2**, 3–16.
102. DEMECZKY, M. and MÁRKUS, P., Hungarian Central Food Research

Institute, Budapest, *Research Report*, 1982. Recent technologies in food processing and their adaptation possibilities. pp. 1–10.
103. ENKINA, L. S., SELYAGIN, V. E., FIKTENGOL'TS, N. N. and CHENSKIH, V. YA., *Khlebopek. Konditer. Prom.*, 1983, **2**, 37–9.
104. ENKINA, L. S., SELYAGIN, V. E., FIKLENGOL'TS, N. N. and CHENSKIH, V. YA., *Khlebopek. Konditer. Prom.*, 1983, **3**, 31–3.
105. MCARTHUR, L. A. and D'APPOLONIA, B. L., *Cereal Chem.*, 1981, **58**, 53–6.
106. SELYAGIN, B. E. and STEPHANOVITS, Z. Z., *Khlebopek. Konditer. Prom.*, 1982, **2**, 32–3.
107. PEI, D. C. T., *Baker's Dig.*, 1982, **56**, 8, 10, 32.
108. HULLS, P. J., *J. Microwave Power*, 1982, **17**, 29–38.
109. ALBRECHT, R. P. and BALDWIN, R. E., *J. Microwave Power*, 1982, **17**, 57–61.
110. MILLER, B. A., KULA, S. A., GOBLE, J. W. and ENOS, H. L., *Poult. Sci.*, 1982, **61**, 463–7.
111. SCHIFFMANN, R. F., MIRMON, A. H., GRILLO, R. J. and BATEY, R. W., USA Patent US 4388335 (1983).
112. MEINASS, U. W., German Federal Republic Patent Application DE 3133030 A 1 (1983).
113. LARSEN, W. L. and DUNCAN, E. R., USA Patent US 4332992 (1982).
114. TANAKA, J., KAI, T. and KURITA, H., USA Patent US 4283614 (1981).
115. MAXTON, K. S., British Patent 1603617 (1981).
116. BOWEN, R. F., FREEDMAN, G. and TEICH, W. W., USA Patent US 4386109 (1983).
117. NODA, T., USA Patent US 4370535 (1983).
118. HORINOUCHI, A., USA Patent US 4356370 (1982).
119. TAKASAKA, A., HORIKOSHI, H. and KASADA, T., USA Patent US 4345132 (1982).
120. BUCK, R. G., USA Patent US 4317977 (1982).
121. DOI, K. and KASHIWAGI, T., USA Patent US 4309585 (1982).
122. NODA, T., USA Patent US 4317976 (1982).
123. KUBIALOVICZ, J. F., USA Patent US 4343978 (1982).
124. ANDRES, C., *Food Process.*, 1981, **42**, 82–3.
125. BENGTSSON, N., *Food Technol.*, 1963, **17**, 97–100.
126. DEMECZKY, M., Hungarian Central Food Research Institute, Budapest, *Élelmiszertudomány*, 1968. **2**, 1–2, 83–85.
127. DEMECZKY, M. and MÁRKUS, P., Hungarian Central Food Research Institute, Budapest, *Research Report*, 1983. Homogeneous defrosting and heating of frozen meat and fruit block by high-frequency fields. pp. 1–19.
128. DEMECZKY, M. and MÁRKUS, P., Hungarian Central Food Research Institute, Budapest, *Research Report*, 1969. Laboratory experiments for the homogeneous heating of frozen egg mixtures. pp. 1–29.
129. DEMECZKY, M., KOPÓCSY, M., MÁRKUS, P. and GROSS, J., Hungarian Central Food Research Institute, Budapest, *Research Report*, 1977. Application of dielectric high-frequency heating in modern salami production for the heating of salami prior to smoking. pp. 1–5.

130. KARABODZHOV, O. and NAUCHNI TR. *Nauchnoizsled. Inst. Konservna Prom.*, *Plovdiv*, 1981, **18,** 45–57.
131. VAVRIK, A., *Prum. Potravin,* 1979, **30,** 413–5.
132. OHLSSON, T., *Livmedelsteknik*, 1980, **22,** 416–8.
133. OTERBAJN, O., *Tehnol. Mesa*, 1980, **21,** 17–8.

Chapter 5

DESIGN AND OPTIMISATION OF FALLING-FILM EVAPORATORS

Mauro Moresi

Dipartimento di Ingegneria Chimica, Materiali, Materie Prime e Metallurgia, Università degli Studi di Roma—La Sapienza, Rome, Italy

SUMMARY

In this chapter the performance and heat transfer capability of falling-film evaporators (FFE) are briefly reviewed with special emphasis on heat and momentum transfer properties and feed and product quality.

A mathematical model of multiple-effect FFE units operating in forward or backward flow was then constructed so as to establish a design strategy aimed at minimising water removal costs without affecting product quality.

In particular, the latter aspect was empirically expressed by using a lumped-parameter (defined as the temperature at which it is possible to hold the liquor examined for a period as long as the typical residence time of MEFFE systems without causing more than 5% loss of its main quality factors) to set up the maximum wall temperature in each FFE.

With specific reference to large scale production of concentrated apple juices, the design strategy based on identical heat transfer surfaces in all the effects was found to be more profitable than that based on identical temperature difference. Moreover, in both cases the optimal number of effects appeared to be independent of the method of feeding, though the reverse flow practice resulted in minimum overall operating costs. However, with forward flow appropriate feed pre-heating seemed to be the only way to reduce the difference in the operating costs of the more economic recirculation six-effect FFE unit operating in backward flow. In this way, it was clearly shown that there is no point in making steam economy prevail over product quality, when experience proves that backward flow is unsuitable for the food product concerned.

NOMENCLATURE

A	Heat transfer surface (m^2)
$\mathbf{A}$	Matrix of coefficients as shown in Table 2
A_{ev}	Heat transfer surface for evaporation (m^2)
A_{ph}	Heat transfer surface for pre-heating (m^2)
a_1, a_2	Parameters defined by eqns (38)–(39)
$\mathbf{B}$	Column vector of known terms
b	Film thickness (m)
C_D	Drag coefficient of falling particles (dimensionless)
C_E	f.o.b. cost of each falling-film evaporator ($)
C_E^*	Base cost for floating-head shell-and-tube exchanger in carbon steel ($)
C_e	Operating costs of the evaporation unit due to electric power consumption ($/h)
C_I	Investment cost of the evaporation unit ($)
C_{Io}	Investment-related costs of the evaporation unit ($/h)
C_{Lo}	Labour costs ($/h)
C_M^*	Base cost for 3600 r/min, totally enclosed, fan-cooled motors ($)
C_0	Overall operating costs of evaporation unit ($/h)
C_P	f.o.b. cost of a generic centrifugal pump ($)
C_P^*	Base cost for one-stage, 3550 r/min, vertically-split case cast-iron pumps ($)
C_{PHE}	f.o.b. cost of a generic plate heat exchanger ($)
C_S	f.o.b. cost of a generic liquid-vapour separator ($)
C_S^*	Base cost of carbon steel drums ($)
C_s	Operating costs of the evaporation unit due to live steam consumption ($/h)
C_{Uo}	Overall utility costs of the evaporation unit ($/h)
C_V	Installed costs of the vacuum system ($)
C_V^*	Base cost of carbon-steel two-stage ejector with barometric pre-condenser and intercondenser ($)
C_w	Operating costs of the evaporation unit due to cooling water consumption ($/h)

c	Local weight flow rate of condensate (kg/s)
c_e	Electric power specific cost ($/kWh)
$c_G, c_p, c_{pa}, c_{wl}, c_{wv}$	Specific heat of non-condensable gases, solution, apple sugar, water and steam (J/(kg K))
c_s	Steam specific cost ($/kg)
c_w	Cooling water specific cost ($/m^3$)
D_S	Inside diameter of a generic liquid–vapour separator (m)
D_{So}	Outside diameter of a generic liquid–vapour separator (m)
df	Degree of freedom (dimensionless)
d_i	Tube inside diameter (m)
d_o	Tube outside diameter (m)
d_p	Average diameter of liquid droplets (m)
E	Recirculation ratio (dimensionless)
E_m	Modulus of elasticity of a generic material of construction (Pa)
$Ei(u)$	Exponential integral function defined by eqn (40)
F	Flow rate of the solution entering each effect (kg/s)
G	Flow rate of non-condensable gases (kg/s)
g	Acceleration due to gravity (m/s^2)
H	Specific enthalpy of vapour phase (J/kg)
H^+	Dimensionless heat transfer coefficient $\{=h_T(\mu^2/\rho^2/g)^{1/3}/k\}$
H_P	Total dynamic head (m)
H_S	Height of a generic liquid–vapour separator (m)
h	Specific enthalpy of liquid phase from each effect (J/kg)
h'	Specific enthalpy of liquid at the top of each evaporator after flashing (J/kg)
$\bar{h}$	Specific enthalpy of liquor entering each effect (J/kg)
h^*	Specific enthalpy of condensed steam or cooling water (J/kg)
h_T	Falling-film heat transfer coefficient (W/(m^2 K))
h_{To}	Condensing film heat-transfer coefficient for external surface (W/(m^2 K))
I_{CEE}	Actual *Chemical Engineering's* Electrical Equipment index

I^*_{CEE}	Reference *Chemical Engineering's* Electrical Equipment index
I_{CEFE}	Actual *Chemical Engineering's* Fabricated Equipment index
I^*_{CEFE}	Reference *Chemical Engineering's* Fabricated Equipment index
I_{CEPC}	Actual *Chemical Engineering's* Pumps and Compressors index
I^*_{CEPC}	Reference *Chemical Engineering's* Pumps and Compressors index
i	Interest rate (dimensionless)
K	Consistency index (Pa s^v)
Ka	Kapitza number $\{=\mu^4 g/(\rho\sigma^3)\}$ (dimensionless)
k	Thermal conductivity of liquor (W/(m K))
k_m	Thermal conductivity of tube material (W/(m K))
L	Overall tube length (m)
l	Generic tube length (m)
l	Dummy index defined by eqns (44)–(45)
M_a, M_w	Molecular weight of air and water (g/mol)
N_e	Number of independent equations (dimensionless)
N_f	Nusselt film thickness parameter $\{=b(\rho^2 g/\mu^2)^{1/3}\}$ (dimensionless)
N_p	Electric power consumption of a generic centrifugal pump (kW)
N_v	Number of independent variables (dimensionless)
n	Number of effects (dimensionless)
n_D	Depreciation time (year)
n_S	Number of shifts per day
n_t	Number of tubes in each evaporator (dimensionless)
P	Overall pressure in each effect (kPa)
P_B	Brake horsepower of a generic centrifugal pump (W)
P_C	Critical pressure (Pa)
P_M	Brake horsepower of the electric motor driving a generic centrifugal pump (kW)
P_m	Motive steam pressure (MPa)
P_T	Closeness of tubes (m)
Pr	Prandtl number $\{=c_p\mu/k\}$ (dimensionless)
p_s	Vapour pressure of solvent (kPa)

Q	Overall heat flow (W)
Q_{ev}	Heat flow for evaporation (W)
Q_{ph}	Heat flow for pre-heating (W)
Q_V	Volumetric flow rate (m³/s)
q	Heat flux (W/m²)
R_d	Combined inside and outside fouling factor (m² K/W)
Re	Reynolds number $\{=4\Gamma/\mu\}$ (dimensionless)
$(Re)_{lw}$	Critical Reynolds number defined by eqn (6) (dimensionless)
Re_p	Particle Reynolds number $\{=\rho_V v_t d_p/\mu_V\}$ (dimensionless)
$(Re)_t$	Critical Reynolds number accounting for the onset of turbulent film flow (dimensionless)
r	Coordinate normal to the wall or tube radius (m)
r^2	Regression coefficient (dimensionless)
S	Flow rate of the liquid phase leaving each effect (kg/s)
s	Local weight flow of evaporating liquor (kg/s)
T	Condensation temperature of the vapour phase (°C)
T_C	Critical temperature (°C)
t	Temperature of the vapour and liquid phases leaving each effect (°C)
$\bar{t}$	Temperature of the liquid phase entering each effect (°C)
t'	Temperature of the liquid phase at the top of each evaporator after flashing (°C)
t_B	Boiling temperature of liquor (°C)
t_{Bw}	Boiling temperature of pure water (°C)
t_p	Temperature of tube wall (°C)
t_S	Shell thickness (m)
t_w	Temperature of cooling water (°C)
U, U'	Local clean heat transfer coefficient (W/(m² K))
U_C	Clean overall heat transfer coefficient (W/(m² K))
U_D	Design overall heat transfer coefficient (W/(m² K))
u	Limit of integration for eqn (40) (dimensionless)
V	Flow rate of vapour leaving each effect (kg/s)
V_E	Motive steam flow rate operating a generic two-stage ejector system (kg/s)

V_P	Steam flow rate required for raw juice pre-heating (kg/s)
v	Local weight flow rate of water evaporated (kg/s)
v_f	Local liquid film velocity (m/s)
$\bar{v}_f$	Average liquid film velocity (m/s)
v_t	Terminal settling velocity (m/s)
W_1	Flow rate of cooling water required at the barometric condenser (kg/s or kg/h)
W_2	Flow rate of cooling water required at the barometric intercondenser (kg/s or kg/h)
w	Local weight flow of condensing steam (kg/s)
X	Solute to water weight ratio in the evaporating liquor (dimensionless)
$\mathbf{X}$	Column vector of unknowns
x	Weight fraction of solute in each S stream (dimensionless or °Brix)
x_G	Weight fraction of the incondensables in the air–vapour mixture leaving the barometric pre-condenser (dimensionless)
y	Weight fraction of solute in the solution entering each effect (dimensionless or °Brix)
y'	Weight fraction of the solute at the top of each evaporator after flashing (dimensionless)
y^+	Universal distance parameter (dimensionless)
z_1	Steam consumption factor, kg of motive steam/kg of air–vapour mixture;
z_2	Non-condensable load factor (dimensionless)
z_3	Steam pressure factor (dimensionless)
α	Thermal diffusivity (m²/s)
β	Thermal loss of each evaporator (dimensionless)
β_D	Cost correction factor for fixed-head exchangers (dimensionless)
β_m	Material cost correction factor (dimensionless)
β_T	Design-type cost factor for one-stage, 1750 r/min, horizontally-split case pumps (dimensionless)
Γ	Weight flow rate per unit channel periphery $\{=S/(\pi d_i n_t)\}$ (kg/(m s))
γ	Shear rate (s⁻¹)
γ_1	Parameter defined by eqn (34) (m² K/W)
γ_2	Parameter defined by eqn (35) (m² K/W)

ΔH	Parameter defined by eqn (62) (J/kg)
ΔT	Temperature difference in each effect (°C)
ΔT_b	Boiling point rise (BPR) (°C)
$\Delta \Theta$	Temperature difference across the liquid film (°C)
$\Delta \tau$	Annual working period of the evaporation unit (h)
δ	Dimensionless film thickness
$\varepsilon, \varepsilon'$	Pre-fixed tolerances for the convergence procedure (dimensionless)
ε_H	Eddy thermal diffusivity (m²/s)
ζ	Variable of integration for eqn (40)
ζ_i	Total module factor of a generic element of the evaporation unit (dimensionless)
η_M	Efficiency of three-phase electric motors (dimensionless)
η_P	Pump efficiency (dimensionless)
Θ	Generic temperature of the liquid falling film (°C)
Θ_i	Temperature at the interface between the liquid falling film and vapour (°C)
λ	Wavelength (m)
λ_w	Latent heat of vaporisation of water at 0°C (J/kg)
μ	Liquor dynamic viscosity (Pa s)
μ_e	Liquor effective viscosity defined by eqn (13) (Pa s)
μ_V	Steam viscosity (Pa s)
μ_w	Water viscosity (Pa s)
v	Flow behaviour index (dimensionless)
v'	Parameter defined by eqn (16) (dimensionless)
ξ	Local weight fraction of solute in the evaporating liquor (dimensionless)
ρ	Liquor density (kg/m³)
ρ_a	'Apple sugar' density (kg/m³)
ρ_m	Material density (kg/m³)
ρ_V	Steam density at the prevailing pressure (kg/m³)
ρ_w	Water density (kg/m³)
σ	Liquor surface tension (N/m)
τ	Shear stress (Pa)
τ_W	Shear stress at the tube wall (Pa)
χ	Dimensionless parameter shown in Table 2
Ψ	Angle of inclination of tubes (degree)
ψ	Size parameter defined by eqn (A.24) (m⁴/s²)

Subscripts

b	Referred to the bottom section of each evaporator
E	Referred to each evaporator
EVB	Referred to an evaporation unit operating in backward feeding
EVF	Referred to an evaporation unit operating in forward feeding
f	Referred to cooling water leaving the barometric condensers
i	Referred to cooling water entering the barometric condensers
j	Referred to a generic effect
K	Expressed in degrees Kelvin
o	Referred to feed liquor and live steam entering the evaporation unit
P	Referred to each centrifugal pump
PHE	Referred to a generic plate heat exchanger
S	Referred to each liquid–vapour separator
t	Referred to the top section of each evaporator

1. INTRODUCTION

Falling-film evaporators have recently found wide application in the food industry because of their peculiar characteristics, such as residence times no longer than a few minutes; high values of the heat transfer coefficient even at low boiling temperatures; minimum loss of the available temperature difference (since the hydrostatic head is negligible and the friction and acceleration effects can be minimised by keeping low the vapour velocity in the cores of the tubes). For this reason, they can safely handle a large variety of thermo-sensitive materials, such as fruit and vegetable juices, milk, whey, beet and cane sugar juices, coffee and tea extracts, etc.

In a previous paper,[1] a mathematical model of multiple-effect falling-film evaporators (MEFFE) was developed by combining the general structure of classic multiple-effect evaporators models with an accurate estimation of the overall heat transfer coefficient (HTC) in each effect based on the assumption that the local clean HTC varied linearly with the temperature of the liquid stream in the pre-heating section of the FFE and exponentially with the solute fraction in the evaporating section of the FFE in question. In this way, by using the correlation of Narayana Murthy and Sarma[2] to predict the clean HTC in falling films, it was possible to obtain a fairly good simulation of the operating variables of several industrial orange and lemon juice double-effect falling-film evaporator plants manufactured by Officine Metalmeccaniche Santoro SpA (Messina, Italy).[1]

The main aim of this chapter is to present the state-of-the-art of such equipment focusing on its description and performance, local and overall heat transfer coefficients, modelling of backward and forward operation, as well as the establishment of an optimal design strategy so as to assess the set of design and operating variables associated with the minimum overall operating costs of the evaporation unit at constant water removal capacity.

Finally, such a design procedure was applied to determine the optimal operating condition for the production of concentrated apple juice.

2. DESCRIPTION AND PERFORMANCE OF FFE

FFE, which have now found wide application in the food industry, were first patented by Kestner in 1899,[3] but only started to be adopted by the fruit juice industry at the end of World War II, as a spin-off from the high-vacuum evaporation techniques specifically refined for the manufacture of perishable materials, such as penicillin.

The first commercial low-temperature orange juice evaporator was installed by Vacuum Foods Company (actually designated as the Foods Division of Coca-Cola Corporation) in 1946.[4] It consisted of a series of vertical stainless steel cylinders about 91·4 cm in diameter equipped with water-jackets. The juice was sprayed against the inside warm surface at the top of the cylinder so that it flowed down by gravity, while the vapour produced was removed by means of a large steam-jet pumping system to maintain an operating temperature of 10–21°C. Although the juice stayed in the evaporator for a long period of time, the concentrate so produced was by far superior to that produced in high temperature evaporators.

Further progress was achieved by the Skinner Company[4] by arranging the evaporators in a shell-and-tube system, so that vapours, drawn off by a steam booster, ascended while the juice descended on the inside surfaces.

Finally, in 1947 a true FFE with both vapours and juice travelling in the same direction was built for the Florida Citrus Canners Cooperative by Majonnier Brothers.[4]

Over the late 40s, the 50s and early 60s a number of companies, such as Buflo-Vak Division of Blaw Knox, Kelly, Gulf Machinery Company, entered the evaporator design and construction field.

Only in 1959 in Florida was the high temperature, short-time (HTST) processing technique applied to fruit juice concentration as a means of improving food quality. Such a HTST evaporator is commonly known as a TASTE (thermally-accelerated short-time evaporation) evaporator. Its basic unit consists of seven stages and four effects plus a flash-cooler and its water removal capacity ranges from 18 000 to 36 000 kg/h. A detailed description of its operation is reported in Refs 4 and 5.

One of the main disadvantages of TASTE evaporators is their relative inflexibility and their high costs. For this reason, other companies, such as A.P.V.–Kestner Ltd (Greenhithe, Kent, UK), Wiegand GmbH (Karlsruhe, FRG), Officine Metalmeccaniche Santoro (Messina, Italy), G. Mazzoni SpA (Busto Arsizio, Italy), etc., entered the evaporator market and were able to manufacture highly flexible FFE so as to treat different kinds of fruit and vegetable juices, as well as other liquid food products.

In particular, highly flexible and reliable six-effect FFE are presently offered so as to minimise evaporation costs.[6]

FFE are usually available in three different versions, that is single-pass FFE (Fig. 1a), recirculation FFE (Fig. 1b) and multi-pass FFE (Fig. 1c). Each single-pass FFE consists of a long vertical one-pass shell-and-tube heat exchanger. Feed enters the top distributor and flows down the tube walls as a thin film by gravity. The heat flux issued from the wall makes a portion of the falling film evaporate, thus producing a liquid–vapour flow which enhances the overall heat transfer coefficient. Then, the liquid–vapour mixture is separated by using a type of cyclone, thus keeping liquid entrainment low even when foaming liquors are processed.

Among all types of evaporators, FFE are characterised by the lowest residence time as a combined result of very thin film flowing at very high velocity and no liquid level in the bottom of either the evaporator or the cyclone separator. In particular, in the latter equipment the liquid level is kept in the down-take leg to prevent the vortex of the cyclone from picking up the liquid from the surface, thus minimising entrainment losses.

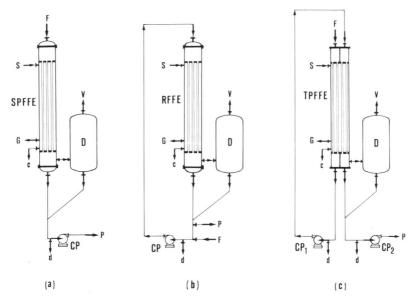

FIG. 1. Typical schemes of single-effect evaporation units consisting of a vertical long-tube FFE, a liquid–vapour separator (D) and a centrifugal pump (CP): (a) single-pass FFE (SPFFE), (b) recirculation FFE (RFFE) and (c) two-pass FFE (TPFFE). Stream identification symbols: c = condensate; d = drainage; F = feed; G = incondensable gases; P = concentrated liquor; S = live steam; V = solvent vaporised.

At low ratios between input flow rate and heat transfer surface or between input and output flow rates, partial drying of tubes may occur, thus reducing the heat transfer effectiveness. This can be avoided by recirculating a portion of the liquid (see Fig. 1b) or adopting multi-pass FFE (Fig. 1c), which obviously involve longer residence times, or using the modified version of the FFE by Vulcan Manufacturing Co. (Cincinnati, USA),[7] which consists of a series of cylindrical shells and truncated cones with a rotating disc located just beneath each cone in order to allow the liquid film to be redistributed more frequently, thus improving the heat and mass transfer capability of the equipment concerned.[7]

One of the main advantages of FFE is that boiling temperatures are practically only affected by boiling point rise on condition that vapour velocity in the tubes is kept lower than 15 m/s.[3]

In this way, heat transfer effectiveness of FFE is maintained even at low temperature differences and/or at low boiling temperatures under

either constant or variable process conditions. In the latter circumstances, such equipment appears to be the only choice among the
large number of evaporator types available.

In fact, at low boiling temperatures, the performance of FFE can be
considered at least good (B level), if not excellent (A level), while that
of the competitive climbing-film evaporators, which is fair-to-good
under steady working conditions, is usually rather poor when the
process conditions are variable for a number of reasons.[7,8]

Referring to severely salting, scaling and/or fouling liquors to be
evaporated, FFE are usually unsuitable because boiling occurs in the
tubes, where most of the deposited solids give rise to a 'sticky' and
non-friable mass.

In particular, salting, scaling and fouling result from the deposition
and growth of various materials on the heat transfer surface, which
decrease the heat transfer rate of the evaporator until a shut-down is
required for cleaning.

Whereas salting results from the deposition of materials, e.g. salts,
with solubility increasing with temperature, scaling is the precipitation
of dissolved substances, or compounds derived from chemical reactions of proteins, sugars and fats, with solubility decreasing with
increasing temperature on the heat transfer surface. Contrary to these,
fouling is due to adhesion of suspended solids present in the feed
liquor (sometimes complicated with protein denaturation in presence
of fibre and pectin,[9] or resulting from corrosion if improper construction materials are used) on the heat transfer surface. In general, the
phenomenon of fouling on heat transfer surfaces is little understood;
for further details Ref. 10 is highly recommended.

Experience shows that fouling is less pronounced at high liquid
velocities; therefore, FFE tend to become taller and taller to increase
weight flow rate per unit width and 10–12 m long tubes have been used
successfully.[6,11]

Although such liquors are better handled in the submerged-tube
forced circulation evaporators than in any other type of evaporators,
in some cases FFE can be more profitably used, especially when mild
salting occurs. In such cases, a certain descaling effect can be obtained
by increasing the feed flow rate or reducing the steam flow rate to
make the liquor unsaturated for a short time without shutting down
the plant. Automatic washing or chemical cleaning can be used to
remove water-soluble or -insoluble deposited solids, thus reducing the
time required to shut down, clean and re-start the evaporation unit.

Although special liquid distributors have been manufactured to make easy the distribution of viscous liquors, the performance of such equipment becomes rather poor as soon as the fluid viscosity is within the range $200^{11}-1000^7$ mPa s.

As far as foaming is concerned, minimum entrainment losses are observed in FFE, since vapour begins to separate from the liquid immediately as it contacts the heat transfer surface and, not explosively, as the liquid enters the liquid–vapour separator.

Finally, single-pass or recirculation FFE can successfully handle thermolabile products with minimum nutritional loss, the mean residence time being less than 10 min.[11,12]

As far as de-watering costs at constant evaporation duty and available overall temperature difference are concerned, maximum steam economy can be obtained with FFE,[8,12] followed by climbing-film evaporators and then by submerged-tube forced-circulation evaporators.[8]

Referring to investment costs, the capital cost of FFE per unit transfer surface is generally higher than that of short or horizontal tube evaporators, but by far smaller than that of forced-circulation evaporators.[7,8,13]

An overall description of the main characteristics of FFE is schematically reported in Table 1.

As far as automation is concerned, FFE, as well as other evaporator types, do not require sophisticated instrumentation. Flow controllers

TABLE 1

PERFORMANCE AND CHARACTERISTICS OF SINGLE-PASS FFE (SPFFE) AND RECIRCULATION FFE (RFFE) AS EXTRACTED FROM REFS 7, 8, 11, 12

Evaporator performance referred to	Main characteristics	SPFFE	RFFE
Heat & momentum transfer properties	Heat transfer coeff. $(kW/m^2/K)$	0·5–2·5	0·5–3
	Liquid residence time	a few s to 4 min	<10 min
	Falling-film velocity	< 20 m/s	<20 m/s
Constant process conditions	High water removal	C to B	A
	Low water removal	A	A
	Low boiling temperature	A	A
	Low temperature difference	A[a]	A[a]

(continued)

TABLE 1—contd.

Evaporator performance referred to	Main characteristics	SPFFE	RFFE
Variable process conditions	Feed	C to B	B to A
	Water removal	C to B	B
	Boiling temperature	B to A	B to A
	Temperature difference	B to A^a	B to A^a
Feed & product quality	Clean	A	A
	Mild salting	B^b	B^b
	Salting	C to B^b	C to B^b
	High salting	D to C^b	D to C^b
	Mild scaling	C to B^c	C to B^c
	Scaling	C^c	C^c
	High scaling	E	E
	Mild fouling	C	C
	Fouling	D to C	D to C
	High fouling	E	E
	Crystallising	E	E
	Erosive suspended solids	B to A	C to B
	Low viscous (1–10 mPa s)	B to A	B to A
	Viscous (10–100 mPa s)	B	B
	High viscous (1–10 Pa s)	D to C	D to C
	Foaming	B	B
	High foaming	B^d	B^d
	Low heat-sensitive	A	B to A
	Heat-sensitive	A	B
	High-heat-sensitive	A	C to B
Installation, operation & maintenance	Main lay-out requirement	headroom	headroom
	Automation required for avoiding	fouling or drying	fouling or drying
	Electric power requirement	a little	a little more
	Mechanical cleaning	not easy	not easy
	Automatic washing or chemical cleaning	suitable	suitable
	Re-start time	immediate	very short

Under the specified circumstances, the FFE performance may be rated as follows:
A = excellent; B = good; C = fair; D = poor; E = unsuitable.

^{a} Provided that the viscosity of liquor is in the range 1–10 mPa s.

^{b} Provided that salting deposits are removed by increasing the feed flow rate or reducing steam flow rate to make the liquid unsaturated for a short time every fortnight or shorter periods.

^{c} Provided that scaling deposits are removed every fortnight or longer periods by diluting liquor for a short time or reducing boiling temperature.

^{d} Provided that proper foam breakers are installed.

on steam and/or feed flow rates, a final concentration controller and an absolute pressure controller on the lower pressure operating effect are usually sufficient to guarantee a safe operation. However, temperature or pressure recorders in each effect can be installed so as to control fouling activity on heat transfer surfaces or localise accidental vacuum losses.

Finally, whatever the number of effects, a single worker per shift is able to effectively control a multiple-effect evaporation unit.

3. HEAT TRANSFER IN FFE

In order to predict the performance of FFE, it is necessary to know the variation in the local heat transfer coefficient (h_T) for non-nucleating conditions (in which all the evaporation from the surface of the film takes place) as a preliminary step to estimate the overall heat transfer coefficient (U_c), which takes into account the variation of the temperature and concentration of the liquor undergoing evaporation over the entire length of the tubes.

In this Section the correlations available in the literature to evaluate h_T are briefly reviewed and, according to the type of flow within the film, used to derive an accurate estimation of U_c.

3.1. Local Heat Transfer Coefficient in Falling-Films

The heat transfer rate in falling films was found to show a large variation because of the different film flow regimes.[14]

Whereas in fluid flow studies it is customary to describe the boundaries between the laminar and turbulent regimes of fluid flow in terms of the Reynolds number (Re) only, the flow regime of a film cannot be uniquely defined as laminar or turbulent owing to the presence of a free surface, which may be smooth or covered with gravity waves or capillary or mixed capillary-gravity waves of various types. Thus, under suitable conditions and at given entrance lengths from the film forming device,[14-16] it is possible to have smooth laminar, wavy-laminar (subdivided into gravity or capillary type) or turbulent flow.

If the flow of a liquid layer, of constant properties and rheological behaviour of the power-law type, down a vertical surface is steady, uniform and two-dimensional (i.e. the flow of a smooth film on an infinitely wide plate or on the internal or external surface of an infinite cylinder outside the acceleration zone), the integration of the equation of motion yields a parabolic profile for the velocity distribution and the weight flow rate per unit channel periphery (Γ) is found to be related to the film thickness (b) as follows[17]

$$\dot{\Gamma} = \frac{\nu}{2\nu + 1} \rho \left(\frac{\rho g}{K}\right)^{1/\nu} b^{(2\nu+1)/\nu} \tag{1}$$

where K is the 'consistency index' and ν is the 'flow behaviour index'. For a Newtonian fluid ν is equal to 1, while K coincides with the fluid dynamic viscosity (μ). In these conditions, eqn (1) reduces to the classic expression of Γ originally developed by Nusselt[18]

$$\Gamma = \frac{\rho^2 g b^3}{3\mu} \tag{2}$$

upon which relies the definition of the dimensionless film thickness (N_f) or Nusselt film thickness parameter

$$N_f = (\tfrac{3}{4} \text{Re})^{1/3} \tag{3}$$

According to the film theory and provided that minimum evaporation takes place at the outer surface of the liquid film, the heat transfer coefficient (h_T) for laminar liquid films can be expressed as

$$h_T = k/b \tag{4}$$

By combining eqns (2) and (4) and introducing the dimensionless heat transfer coefficient (H^+) and Reynolds number (Re), it is possible to derive the Nusselt model[19]

$$H^+ = 1 \cdot 1 \, \text{Re}^{-1/3} \tag{5}$$

However, such an equation can be used to predict heat transfer coefficients only under true laminar film flow, that is for Re less than a critical value $(\text{Re})_{lw}$ accounting for the first formation of waves.[20]

Among the number of theories covering the onset of instability in film flow on the outer surface of a vertical surface (see also Ref. 18), the Kapitza theory indicates that the capillary waves would form on the laminar–layer surface for Re exceeding the value

$$(\text{Re})_{lw} = 2 \cdot 43 \, \text{Ka}^{-1/11} \tag{6}$$

where Ka is the Kapitza number. Such a theory, assuming that wavelengths (λ) are long compared to film thickness ($\lambda/b \geqslant 13 \cdot 7$), indicated that the average film thickness for a given flow rate is reduced by the action of the ripples

$$N_f = 1 \cdot 22 \, Re^{0 \cdot 22} \qquad (7)$$

In the circumstances, higher dimensionless heat transfer coefficients than those observed under smooth laminar flow were experienced. To obtain a better reconstruction of the experimental results, Chun and Seban[19] replaced eqn (5) with the following empirical relationship

$$H^+ = 0 \cdot 822 \, Re^{-0 \cdot 22} \qquad (8)$$

This contrasts with Dukler and Bergelin's[21] previous statements that at Reynolds numbers below 1000 the Nusselt equation (3) predicted the experimental mean film thickness even when waves were present on the outer surface, or with Sarma et al.'s[22] theoretical demonstration that in the region of evaporation and for Re < 1000 the asymptotic dimensionless heat transfer coefficient H^+ is a different function of Re

$$H^+ = 1 \cdot 66 \, Re^{-0 \cdot 31} \qquad (9)$$

There are several reports in the literature of the critical Reynolds number, $(Re)_t$, at which turbulent film flow commences. The bulk of the evidence seems to support a lower value in the region 1000–1600, with a less well-marked upper value of about 3200.[18] Such scattering in the experimental values of $(Re)_t$ can be explained by pointing out that the transition to turbulence in a thin film is a gradual process, thus making it practically impossible to define a single, sharp, critical Reynolds number.[23]

In the turbulent region (Re > 1600), the film thickness can be roughly estimated as follows[18]

$$N_f = 0 \cdot 266 \, Re^{1/2} \qquad (10)$$

In such a flow regime the film structure essentially consists of two portions: a base-film portion flowing next to the wall (the so-called 'laminar sublayer') and turbulent waves. Such waves appear to have an extremely large base-to-height ratio and move independently of each other, thus being better envisioned as a train of single segments of thicker turbulent films. Since each wave moves at a much higher velocity than the base film, it continuously overtakes the fluid in the base-film found in front of it and mixes with it so that the base film loses its identity beneath the wave. Contemporaneously, an equivalent amount of fluid is left behind and undergoes relaminarisation to make up the laminar base-film at the back of the waves.

For the turbulent flow conditions of the liquid film, heat transfer is controlled by both thermal (α) and eddy (ε_H) diffusivity

$$q = \rho c_p (\alpha + \varepsilon_H) \frac{d\theta}{dr} = h_T \, \Delta\theta \tag{11}$$

where $d\theta/dr$ is the temperature gradient along the coordinate normal to the wall, θ is the generic temperature, and $\Delta\theta$ the temperature difference across the liquid film. Integration of eqn (11) over the overall film thickness gives the general expression for the heat transfer coefficient, which, unfortunately, is a complex function of both eddy diffusivity of momentum and heat, as well as velocity distribution. References 23–27 illustrate the resolution of the aforementioned integral expression of h_T for a film of uniform thickness by assuming different turbulent transport mechanisms.

Table 2 reviews the main equations reported in the literature for heat transfer rates in turbulent falling film.[2,18,28–32] In a previous paper[1] the reliability of the only equations which take into account the effect

TABLE 2

AVAILABLE EQUATIONS FOR THE PREDICTION OF HEAT TRANSFER RATE IN TURBULENT FALLING FILMS[1]

No.	Equation	Flow characteristics	Ref.
1	$H^+ = 0\cdot01\,(\mathrm{Re}\,\mathrm{Pr})^{1/3}$	Turbulent motion down the inner walls of pipes	28
2	$H^+ = 0\cdot02\,\mathrm{Re}^{1/3}\,(\sin\Psi)^{0\cdot2}$	Turbulent motion down flat plates with different angles (Ψ) of inclination	29
3	$H^+ = 8\cdot7 \times 10^{-3}\,\mathrm{Re}^{0\cdot4}\,\mathrm{Pr}^{0\cdot34}$	Turbulent motion outside of a metal rod heated internally by hot water	30
4	$H^+ = 6\cdot92 \times 10^{-3}\,\mathrm{Re}^{0\cdot345}\,\mathrm{Pr}^{0\cdot4}$	Turbulent motion down the inside surface of tubes	31
5	$H^+ = 8\cdot54 \times 10^{-4}\,\mathrm{Re}^{0\cdot65}$	Turbulent motion of water only down the inner walls of pipes	32
6	$H^+ = 3\cdot8 \times 10^{-3}\,\mathrm{Re}^{0\cdot4}\,\mathrm{Pr}^{0\cdot65}$	Turbulent flow down the surface of an electrically heated vertical tube	19
7	$H^+ = 0\cdot89\delta^{1/3}/\chi$ $\chi = 5 + \{\tan^{-1}(2\cdot73\sqrt{\mathrm{Pr}})$ $- \tan^{-1}(0\cdot455\sqrt{\mathrm{Pr}})\}/$ $(0\cdot091\sqrt{\mathrm{Pr}}) + \ln(\delta/30)/$ $(0\cdot36\,\mathrm{Pr})$	Theoretical expression applicable for $\mathrm{Pr} > 1$ and $\delta > 30$	2

Courtesy of Journal of Food Technology.

of the Prandtl number (Pr) and refer to the turbulent flow down tubes and pipes (i.e. the correlations developed by McAdams,[28] Wilke,[30] Ahmed and Kaparthi,[31] Chun and Seban[19] and Narayana Murthy and Sarma[2]) was examined by testing their capability at analysing the performance of a few industrial orange and lemon juice double-effect falling-film evaporators manufactured by Officine Metalmeccaniche Santoro SpA (Messina, Italy). Since the best simulation of the operating variables of the aforementioned FFE was obtained by using the correlation of Narayana Murthy and Sarma,[2] such an equation is recommended for the estimation of H^+ under turbulent flow conditions.[1] Moreover, the correlation concerned (Table 2) is based on an explicit form of the dimensionless film thickness (δ), which originally applied for values of the universal distance parameter $y^+ > 30$ (that is, in the turbulent zone of the von Kármán theory[33]), while in Ref. 1 it was empirically extended to the buffer layer ($5 < y^+ \leqslant 30$) by newly correlating the data of Dukler and Bergelin[21] and Belkin et al.[34] in the following manner

$$\ln \delta = 0.786 + 0.103 \ln Re + 0.41 (\ln Re)^2 \quad \text{for} \quad \delta > 5 \quad (12)$$

where δ is the dimensionless film thickness for Reynolds numbers exceeding 40.

In conclusion, for the sake of simplicity the dimensionless film thickness (N_f) was expressed by using eqn (3) for $Re < 1600$ or eqn (10) for $Re > 1600$, whereas the dimensionless heat transfer coefficient (H^+) was based upon the Nusselt equation (5) for $Re < (Re)_{lw}$, upon Chun and Seban's[19] equation (8) for $(Re)_{lw} < Re < (Re)_t$ and finally upon the correlation of Narayana Murthy and Sarma[2] (equation No. 7 in Table 2) for $Re > (Re)_t$, where $(Re)_t$ results from the intersection of the last two equations to use. In this way, neither N_f or H^+ shows any discontinuity when plotted against the Reynolds number.

Since a great amount of liquid foods undergoing concentration by evaporation exhibit pseudoplastic (shear-thinning) behaviour[35] and no experimental data on heat transfer to non-Newtonian falling liquid films are actually available in the literature, the most straightforward way to deal with such a problem would appear to replace the dynamic viscosity (μ) of Newtonian fluids with an effective viscosity (μ_e) defined as that viscosity which makes the Nusselt equation (2) fit any set of laminar flow conditions for time-independent fluids in almost the same way traditionally used to define the generalised Reynolds number for tube flow[17]

$$\mu_c = \frac{g\rho b}{4(\bar{v}_f/b)} = \frac{\tau_w}{3(\bar{v}_f/b)}$$ (13)

where τ_w is the shear stress at the tube wall.

Under the hypothesis of film thickness far smaller than the inside tube diameter ($b \ll d_i/2$) and following the procedure used by Rabinowitsch and Mooney[17] to construct a general equation relating the average fluid velocity ($\bar{v}_f$) and τ_w for laminar flow of time-independent fluids in cylindrical tubes, it was possible to establish a similar relationship between $\bar{v}_f$ and τ_w for laminar falling films

$$\frac{2v+1}{v+1}\left(\frac{\bar{v}_f}{b}\right) = \frac{1}{\tau_w}\int_0^{\tau_w}\left(-\frac{dv_f}{dr}\right)d\tau_{rl}$$ (14)

thus making it possible to evaluate the shear rate at the wall

$$\left(-\frac{dv_f}{dr}\right)_w = \frac{2v+1}{v+1}\left(\frac{v'+1}{v'}\right)\left(\frac{\bar{v}_f}{b}\right)$$ (15)

with

$$v' = d \ln \tau_w/d \ln (\bar{v}_f/b)$$ (16)

where v' coincides with v for power–law fluids.

For the above category of non-Newtonian fluids the substitution of eqn (15) into eqn (13) yields the following

$$\mu_c = \frac{K}{3}\left(\frac{2v+1}{v}\right)^v\left(\frac{\bar{v}_f}{b}\right)^{v-1}$$ (17)

with

$$\bar{v}_f = \Gamma/(\rho b)$$ (18)

Obviously, eqn (17) reduces to the dynamic viscosity (μ) of Newtonian fluids when v is equal to 1.

In this way, by using a trial and check procedure it is possible to estimate both film thickness and heat transfer coefficient in non-Newtonian falling-films whatever the prevailing flow conditions.

3.2. Calculation of the Overall Heat Transfer Coefficient in FFE

As the liquor to be evaporated flows down the tube walls of the FFE, both its temperature and concentration increase, thus affecting the value of the local heat transfer coefficient (h_T).

From a design point of view it is easier to define an overall heat transfer coefficient referred to the difference between the condensation temperature (T) of steam and boiling temperature (t) of liquor leaving the evaporator:

$$U_C = \frac{Q}{A(T - t)} \qquad (19)$$

where Q is the overall heat flow and A is the outside heat transfer surface of the FFE examined.

Generally speaking, the inlet temperature $(\bar{t})$ of the liquid phase (F) at the top of the evaporator may or may not coincide with the equilibrium boiling temperature (t_B) at the prevailing pressure P at the exit section of the evaporator (for the sake of simplicity the pressure drop through the heat exchanger was assumed to be negligible).

For $t > t_B$ F is adiabatically flashed, thus varying its concentration and temperature from y to y' and from $\bar{t}$ to t', respectively (see step 16 of the Procedure of Calculation described in Section 4).

For $t < t_B$ the heat flux issued from the wall will make the falling film pre-heat up to the bubble point before producing a liquid–vapour flow.

In the pre-heating section of the FFE (Fig. 2a) the energy balance referred to the differential volume of FFE and heat flow across the tube wall yield the following

$$dw\{H(T) - h^*(T)\} = F\,dh = U(T - \theta_i)\,dA' = dQ_{ph} \qquad (20)$$

where $H(T)$ and $h^*(T)$ are, respectively, the specific enthalpy of steam and condensed steam; $h(\theta, x)$ is the specific enthalpy of liquor, which is a function of temperature (θ) and solute fraction (x); A' is the outside heat transfer surface of FFE where falling film pre-heating occurs; U is the local clean HTC referred to the external surface of FFE, that is

$$\frac{1}{U} = \frac{1}{h_T(d_i/d_o)} + \frac{d_o}{2k_m}\ln(d_o/d_i) + \frac{1}{h_{To}} \qquad (21)$$

where k_m is the thermal conductivity of the tube material, h_{To} is the condensate film HTC on the steam side referred to the outside surface and evaluated in accordance with Nusselt's theory;[36] h_T is the falling film HTC calculable as shown in Section 3.1 and d_i and d_o are the inside and outside diameters of tubes, respectively.

By integrating eqn (20), it is possible to evaluate the heat transfer

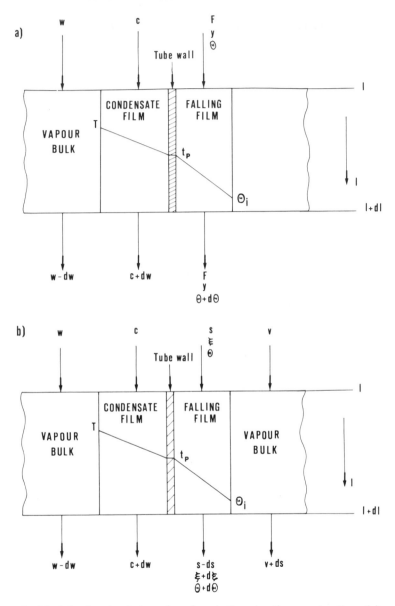

FIG. 2. Models for heat transfer description in the pre-heating (a) and
evaporating (b) sections of FFE.

surface (A_{ph}) required to pre-heat the liquor up to t_B

$$A_{\text{ph}} = F \int_t^{t_B} \frac{(\partial h / \partial \theta)_{\xi = y}\, d\theta}{U(T - \theta_i)} \tag{22}$$

while the corresponding heat flow (Q_{ph}) is

$$Q_{\text{ph}} = F\{h(t_B) - h(\bar{t})\} \tag{23}$$

with

$$t_B = t_{Bw}(P) + \Delta T_b(P, y) \tag{24}$$

where $t_{Bw}(P)$ and t_B are, respectively, the boiling points of pure water and liquor with a solute fraction y at the prevailing pressure P.

As soon as the temperature of liquor is equal to, or greater than, t_B, a different heat transfer mechanism has to be taken into account.

With reference to Fig. 2b, the solute material balance, heat balance across the differential volume and total heat flow across the tube wall can be written as follows

$$\xi\, ds = s\, d\xi \tag{25}$$

$$dw\{H(T) - h^*(T)\} = \left\{ s\frac{\partial h}{\partial \theta} + v\frac{\partial H}{\partial \theta} \right\} d\theta + \{H(\theta) - h(\theta, \xi)\}\, ds$$

$$= U(T - \theta_i)\, dA'' = dQ_{ev} \tag{26}$$

while the overall- and solute-material balances from $A = A_{\text{ph}}$ to a generic $A > A_{\text{ph}}$ involve

$$F = s + v \tag{27}$$

and

$$Fy = s\xi \tag{28}$$

As the falling film evaporates, the solute fraction (ξ) in the liquor increases, thus greatly affecting the physical properties of the liquid phase.

Since the falling film generally exerts the controlling resistance to heat transfer, the local clean $HTC(U)$ results in being a complex function of the solute fraction—that is, $U(\xi)$—the boiling temperature of liquor being dependent on solute concentration only {see eqn (24)}.

By combining eqns (25)–(28) and integrating modified eqn (26) between the inlet (y) and outlet (x) solute fractions of liquor in the evaporator, it was possible to evaluate the heat transfer surface (A_{ev}) required to concentrate the liquor between the aforementioned limits, that is

$$A_{ev} = Fy \int_y^x \left\{ \left[\frac{\partial h}{\partial \theta} + \left(\frac{\xi}{y} - 1 \right) \frac{dH}{d\theta} \right] \xi \frac{\partial \Delta T_b}{\partial \xi} + (H - h) \right\} \frac{d\xi}{\xi^2 U(\xi)(T - \theta_i)}$$

with   (29)

$$\theta_i = t_{Bw}(P) + \Delta T_b(P, \xi) \tag{30}$$

while the associate heat flow is

$$Q_{ev} = F \left\{ [H(t) - h(t_B, y)] - \frac{y}{x} [H(t) - h(t)] \right\} \tag{31}$$

By summing up the heat transfer surfaces (A_{ph} and A_{ev}) and heat flow (Q_{ph} and Q_{ev}) concerning the pre-heating and evaporating sections of the FFE examined, it is possible to derive the following expression for the reciprocal of U_C from eqns (19), (22), (23), (29) and (31)

$$\frac{1}{U_c} = \frac{(A_{ph} + A_{ev})(T - t)}{Q_{ph} + Q_{ev}} = \frac{(T - t)}{\{H(t) - h(\bar{t})\} - \frac{y}{x} \{H(t) - h(t)\}}$$

$$\times \left\{ \left[\int_{\bar{t}}^{t_B} \frac{(\partial h / \partial \theta)_{\xi = y}}{U(T - \theta_i)} d\theta \right. \right.$$

$$\left. \left. + y \int_y^x \frac{[\partial h / \partial \theta + (\xi / y - 1)(dH / d\theta)] \xi (\partial \Delta T_b / \partial \xi) + (H - h)}{\xi^2 U(\xi)(T - t_{Bw} - \Delta T_b)} d\xi \right\} \right. \tag{32}$$

The application of eqn (32) appears to be troublesome, especially if it is inserted into an iterative routine, such as that concerning the design of multiple-effect FFE.

By assuming that t_B approximately coincides with t_{Bw} (ΔT_b being far smaller than t_{Bw}), thus extrapolating that in these circumstances the derivative of ΔT_b with respect to ξ can be practically considered as equal to zero, that the specific enthalpies of liquor and vapour phases are constant over the entire length of FFE as a result of constant pressure in each evaporator, and that θ_i practically coincides with θ owing to the narrowness of the falling film, eqn (32) can be transformed into

$$\frac{1}{U_C} = \gamma_1 + \gamma_2 \tag{33}$$

with

$$\gamma_1 = \frac{(T - t)(\partial h/\partial t)_{\xi=y}}{\{H(t) - h(\bar{t})\} - \dfrac{y}{x}\{H(t) - h(t)\}} \int_{\bar{t}}^{t_B} \frac{d\theta}{U(\theta)(T - \theta)} \quad (34)$$

and

$$\gamma_2 = \frac{y}{\{H(t) - h(\bar{t})\}/\{H(t) - h(t)\} - y/x} \int_y^x \frac{d\xi}{\xi^2 U(\xi)} \quad (35)$$

where the first term (γ_1) on the right-hand side of eqn (33) refers to the eventual pre-heating section of FFE, which is omitted if $\bar{t}$ is equal to, or greater than, t_B; while the second one (γ_2) refers to the evaporating section of FFE.

By assuming that the local clean HTC, respectively, varies linearly with the temperature of the liquid stream and exponentially with the solute fraction in the pre-heating and evaporating sections of the FFE, it was possible to derive the following analytical expressions of γ_1 and γ_2[1]

$$\gamma_1 = \frac{(\partial h/\partial t)_{\xi=y}(T - t)(t_B - \bar{t})}{\{H(t) - h(\bar{t})\} - \dfrac{y}{x}\{H(t) - h(t)\}} \cdot \frac{\ln\{U(t_B)(T - \bar{t})/U(\bar{t})/(T - t_B)\}}{U(t_B)(T - \bar{t}) - U(\bar{t})(T - t_B)}$$

$$(36)$$

and

$$\gamma_2 = \frac{y}{\{H(t) - h(t)\}/\{H(t) - h(t)\} - y/x}$$

$$\times \left\{\frac{a_1}{a_2}[Ei(a_1 x) - Ei(a_2 y)] + \frac{1}{yU(y)} - \frac{1}{xU(x)}\right\} \quad (37)$$

with

$$a_1 = \ln\{U(y)/U(x)\}/(x - y) \quad (38)$$

$$a_2 = \{U(y)\}^{x/(x-y)}/\{U(x)\}^{y/(x-y)} \quad (39)$$

and

$$Ei(u) = \int_{-\infty}^u \frac{e^\zeta}{\zeta}\,d\zeta = 0 \cdot 5772157 + \ln|u| + \sum_1^\infty \frac{u^k}{kk!} \quad (40)$$

Having previously[1] verified that the above assumptions exhibited a maximum deviation (about 20%) between the rigorous and approximated values of U_C in good agreement with the mean standard error

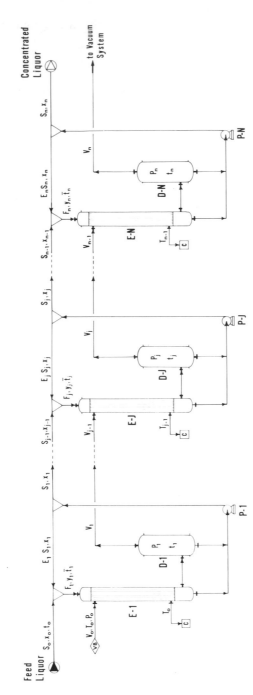

Fig. 3. Typical flow diagram of a forward MEFFE system. Equipment identification items: D = liquid–vapour separator; E = FFE; P = centrifugal pump. Utility identification items: c = cooling water; VB = low pressure steam. Stream symbols as in the Nomenclature section. (Courtesy of Journal of Food Technology).

of a large number of heat transfer correlations, it would seem to be sufficiently safe and simple to design and analyse multiple-effect FFE by replacing eqn (32) with the set of eqns (33) and (36–40), as will be shown below.

4. MODELLING OF MULTIPLE-EFFECT FFE

4.1. Mathematical Model

Modelling of multiple-effect FFE (MEFFE) operating in either forward or backward flow was carried out by taking into account the following.

(1) At the prevailing conditions the solute can be regarded as non-volatile, thus allowing a complete vaporisation of the solvent without removal of the solute.

(2) The available temperature difference in each effect is reduced by the effect of boiling point rise only.

(3) Incondensable gases (such as, air leakage and air release from the feed liquors) do not affect the overall pressure and the effectiveness of heat transfer in the system, whereas their amount has to be estimated in order to design the vacuum equipment (e.g. ejector or vacuum pump, barometric or surface condensers, extraction pump or barometric leg), as shown in Section 8.

With reference to the typical schemes of forward (Fig. 3) and backward (Fig. 4) MEFFE systems, it is possible to describe both systems by imposing the overall and solute material balances, and the heat balance across each generic effect j:

$$S_{l-1} = S_l + V_j \qquad (41)$$

$$S_{l-1}x_{l-1} = S_l x_l \qquad (42)$$

$$S_{l-1}h_{l-1} + V_{j-1}(H_{j-1} - h_{j-1})(1 - \beta_j) = S_l h_l + V_j H_j \qquad (43)$$

where all the symbols are reported in the Nomenclature section. In particular, the dummy index l is equal to

$$l = j \qquad \text{in forward flow} \qquad (44)$$

$$l = n + 1 - j \qquad \text{in backward flow} \qquad (45)$$

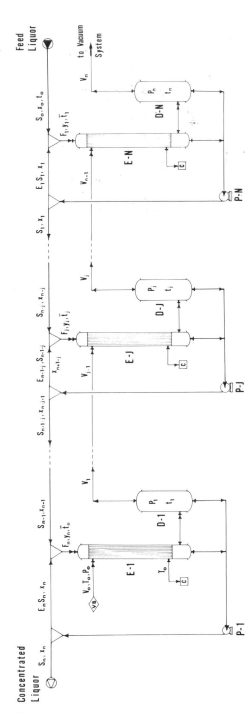

Fig. 4. Typical flow diagram of a backward MEFFE. Equipment and utility identification items and stream symbols as in Fig. 3.

The performance of each FFE is also described by the following heat transfer equation

$$V_{j-1}(H_{j-1} - h^*_{j-1})(1 - \beta_j) = U_{Dj}A_j(T_{j-1} - t_j) \qquad (46)$$

where T_{j-1} is the condensation temperature of the steam, which comes from the $(j-1)$th effect and condenses in the shell of the j-th evaporator

$$T_{j-1} = t_{j-1} - \Delta T_{b,j-1} \quad \text{for} \quad j = 2, 3, \ldots, n \qquad (47)$$

and T_0 is the condensation temperature of live steam.

Provided that the equilibrium conditions are achieved in each effect, the temperature of the boiling liquor and solvent vaporised are equal (t_j). Moreover, owing to the boiling point elevation $(\Delta T_{b,j})$ of the liquor, the vapour leaving the j-th effect is superheated, the overall pressure (P_j) in the j-th separator being equal to vapour pressure of the solvent

$$P_j = p_s(T_j) \qquad (48)$$

where p_s is the vapour pressure of solvent.

When the liquor recirculates through the evaporator, the weight flow rate (F_j), composition (y_j) and temperature $(\bar{t}_j)$ of the solution entering the j-th evaporator can be easily determined by solving the following overall- and solute-material and heat balances

$$F_j = S_{l-1} + E_l S_l \qquad (49)$$

$$F_j y_j = S_{l-1} x_{l-1} + E_l S_l x_l \qquad (50)$$

$$F_j \bar{h}_j = S_{l-1} h_{l-1} + E_l S_l h_l \qquad (51)$$

where E_l is the recirculation ratio. When single pass FFE are considered E_l is equal to zero.

4.2. Procedure of Calculation
The system of non-linear equations (41–51) can be used to design MEFFE on condition that the number of degrees of freedom (df) concerning this problem is obviously zero. With reference to Figs 3 & 4, the independent variables are

$$S_j; x_j; t_j; V_j; P_j; T_j \quad \text{for} \quad j = 0, 1, \ldots, n$$

and

$$A_j; U_{Dj}; \beta_j; E_j; F_j; y_j; \bar{t}_j \quad \text{for} \quad j = 1, 2, \ldots, n$$

and their overall number is

$$N_v = 13n + 6$$

whereas the number of independent equations (N_e) is

$$N_e = 9n$$

Thus, the df of the system is

$$df = N_v - N_e = 4n + 6$$

For practical purposes the following parameters are usually specified to make the problem determinate: weight flow rate (S_0), composition (x_0) and temperature (t_0) of the feed liquor; solute composition of the final product (x_n); operating conditions (T_0, P_0, P_n); and, finally, thermal loss (β_j), overall heat transfer coefficient (U_{Dj}) and recirculation ratio (E_j) for each effect.

However, at this stage the problem is still indeterminate with $(n-1)$ degrees of freedom and a determinate solution can be obtained as follows.

(1)   Identical heat transfer surfaces in all the evaporators

$$A_j = A = \text{const} \quad \text{for} \quad j = 1, 2, \ldots, n \qquad (52)$$

This assumption gives the advantage of designing a single evaporator whatever the number of effects, thus reducing the engineering, construction and installation costs of the evaporation unit.

(2)   Different heat transfer surfaces in all the evaporators

$$A_j \neq A_{j+1} \quad \text{for} \quad j = 1, 2, \ldots, n \qquad (53)$$

This working hypothesis has to be coupled with some other condition being a problem of optimisation as indicated by Holland.[37] For instance, the required $(n-1)$ values of A_j might be assigned by minimising the overall costs of the evaporation unit or assuming identical temperature differences in all the evaporators

$$T_{j-1} - t_j = \Delta T = \text{const} \quad \text{for} \quad j = 1, 2, \ldots, n \qquad (54)$$

All the mentioned assumptions will be further analysed (see Section 6).

Once given the mentioned process and design parameters, the number of effects (n) and some geometrical parameter (size—d_o, d_i,

L—and closeness—P_T—of tubes and their type of pitch), the design procedure of MEFFE may be summarised as follows.

(1) Calculate the first trial values of solvent flow rate vaporised in each effect

$$V_j = S_0(1 - x_0/x_n)/n \quad \text{for} \quad j = 1, 2, \ldots, n \tag{55}$$

(2) Evaluate the liquor flow rate (S_l) in each stage by using eqn (41) for l ranging from 1 to n.

(3) Estimate the solute weight fraction (x_l) in each stage by means of eqn (42) for l varying from 1 to n.

(4) Calculate the condensation temperature of the solvent vaporised in the n-th effect from eqn (48)

$$P_n = p_s(T_n) \tag{56}$$

(5) Assume a tentative set of t_j values for all the n effects.

(6) Calculate the boiling point rise $(\Delta T_{b,j})$ of liquor in each stage. This parameter is a function of temperature (t_j) and solute concentration (x_j)

$$\Delta T_{b,j} = f(t_j, x_j) \tag{57}$$

Unfortunately, this function is available only for a limited number of liquors: for instance, lemon,[38] orange,[39] apple[40] and grape[41] juices.

(7) Calculate the specific enthalpy of each stream in each effect

$$h_j = c_p t_j \tag{58}$$

$$h_j^* = c_{wl} T_j \tag{59}$$

$$H_j = \lambda_w + c_{wv} T_j \tag{60}$$

where T_j can be evaluated by using eqn (47), while the reference temperature for enthalpy calculation is 0°C.

(8) Calculate the live steam flow rate (V_0) from eqn (43) by assuming $j = 1$.

(9) Estimate the condensation temperatures $(T_1, T_2, \ldots, T_{n-1})$ of solvent vaporised in all the effects, except the last one evaluated at step 4, by solving the following system of $(n-1)$ linear equations (obtained by substituting eqn (47) into eqn (46) and dividing by $U_{Dj}A_j$)

$$V_{j-1} \Delta H_{j-1}/(U_{Dj}A_j) = T_{j-1} - T_j - \Delta T_{b,j} \tag{61}$$

where

$$\Delta H_{j-1} = (H_{j-1} - h_{j-1}^*)(1 - \beta_j) \tag{62}$$

If the heat transfer surfaces are equal—see eqn (52)—A can be evaluated by summing up both sides of eqn (61) for j ranging from 1 to n, thus yielding

$$A = \frac{\sum_{1}^{n} V_{j-1}\,\Delta H_{j-1}/U_{Dj}}{T_0 - T_n - \sum_{1}^{n}\Delta T_{b,j}} \tag{63}$$

If all the heat transfer surfaces are different—see eqn (53)—$(n-1)$ values of A_j are to be prefixed in a certain way, while the remaining one (for instance, A_1) can be calculated by applying the same procedure used to derive eqn (63)

$$A_1 = \frac{V_0\,\Delta H_0/U_{D1}}{T_0 - T_n - \sum_{1}^{n}\Delta T_{bj} - \sum_{2}^{n} V_{j-1}\,\Delta H_{j-1}/(A_j U_{dj})} \tag{64}$$

In both cases the condensation temperatures (T_j) can be easily determined by solving eqn (61) for j varying sequentially from 1 to $(n-1)$. If the temperature differences in all the evaporators are equal—see eqn (54)—ΔT can be calculated by substituting eqn (47) into the design condition (54) and adding up its left- and right-hand members for $j = 1, 2, \ldots, n$

$$\Delta T = \left(T_0 - T_n - \sum_{1}^{n}\Delta T_{b,j}\right)\bigg/ n \tag{65}$$

Then, the condensation temperatures (T_j) can be determined one by one, by solving in sequence eqn (54)

$$T_j = T_{j-1} - \Delta T_{b,j} - \Delta T \quad \text{for} \quad j = 1, 2, \ldots, n \tag{66}$$

Finally, each heat transfer surface (A_j) can be calculated by means of eqn (46) for $j = 1, 2, \ldots, n$.

(10) Calculate the boiling temperature (t_j) of the liquor in each effect from eqn (47) for $j = 2, 3, \ldots, n$.

(11) Evaluate the specific enthalpies of all the liquid and vapour streams by using eqns (58–60) for j ranging from 1 to $(n-1)$.

(12) Determine the weight flow rates of all the vapour and liquid streams, except those of the feed (S_0) and final (S_n) liquors, by solving the $2n$ linear equation—(43) and (41)—system, by using for instance the method of Gaussian elimination.[42] Such a system can be written in matrix form as

$$\mathbf{AX} = \mathbf{B} \tag{67}$$

with

$$\mathbf{A}(2n, 2n) = \begin{vmatrix} \mathbf{A}_{11}(n, n+1) & \mathbf{A}_{12}(n, n-1) \\ \mathbf{A}_{21}(n, n+1) & \mathbf{A}_{22}(n, n-1) \end{vmatrix}$$

$$\mathbf{X}(2n) = \begin{vmatrix} \mathbf{X}_1(n+1) \\ \mathbf{X}_2(n-1) \end{vmatrix} = \begin{vmatrix} V_0 \\ V_1 \\ \cdot \\ \cdot \\ \cdot \\ V_n \\ S_1 \\ \cdot \\ \cdot \\ S_{n-1} \end{vmatrix}$$

and

$$\mathbf{B}(2n) = \begin{vmatrix} \mathbf{B}_1(n) \\ \mathbf{B}_2(n) \end{vmatrix} = \begin{vmatrix} -h_0 S_0 \\ 0 \\ \cdot \\ \cdot \\ \cdot \\ h_n S_n \\ S_0 \\ 0 \\ \cdot \\ -S_n \end{vmatrix}$$

While the rectangular partition matrices $\mathbf{A}_{12}(n, n-1)$ and $\mathbf{A}_{22}(n, n-1)$ and the column vectors $\mathbf{X}$ and $\mathbf{B}$ have the same structure whatever the MEFFE system examined, the elements of the remaining partition matrices $\mathbf{A}_{11}(n, n+1)$ and $\mathbf{A}_{21}(n, n+1)$ assume different values according to backward or forward flow (see Table 3).

(13) Estimate a new set of solute weight fractions (x_l') by using eqn (42) for $l = 1, 2, . . , (n-1)$.

(14) Check if the difference between the old (x_j) and new (x_j') sets of solute weight fractions in each effect is less than a prescribed tolerance (ε)

$$|x_j' - x_j| < \varepsilon \qquad j = 1, 2, \ldots, (n-1) \qquad (68)$$

If the above difference is greater than ε, the calculation procedure is repeated from step 6 by using the method of direct substitution modified to speed the convergence. If condition (68) is satisfied for $j = 1, 2, \ldots, (n-1)$, the iterative process is said to have converged and the calculation procedure can continue.

TABLE 3

STRUCTURE OF THE FOUR PARTITION MATRICES EXTRACTED FROM THE MATRIX OF COEFFICIENT **A** OF eqn (67) AS A FUNCTION OF BACKWARD OR FORWARD FLOW

Partition matrix	Backward flow	Forward flow
$A_{11}(n, n+1)$	$\begin{vmatrix} 0 & 0 & \cdots & 0 & \Delta H_{n-1} & -H_n \\ 0 & 0 & \cdots & \Delta H_{n-2} & -H_{n-1} & 0 \\ \cdot & & \cdots & & \cdot & \cdot \\ 0 & \Delta H_1 & \cdots & 0 & 0 & 0 \\ \Delta H_0 & -H_1 & \cdots & 0 & 0 & 0 \end{vmatrix}$	$\begin{vmatrix} \Delta H_0 & -H_1 & \cdots & 0 & 0 & 0 \\ 0 & \Delta H_1 & \cdots & 0 & 0 & 0 \\ \cdot & & \cdots & & & \cdot \\ 0 & 0 & \cdots & \Delta H_{n-2} & -H_{n-1} & 0 \\ 0 & 0 & \cdots & 0 & \Delta H_{n-1} & -H_n \end{vmatrix}$
$A_{12}(n, n-1)$		$\begin{vmatrix} -h_1 & 0 & \cdots & 0 & 0 & 0 \\ h_1 & -h_2 & \cdots & 0 & 0 & 0 \\ \cdot & & \cdots & & & \cdot \\ 0 & 0 & \cdots & 0 & h_{n-2} & -h_{n-1} \\ 0 & 0 & \cdots & 0 & 0 & h_{n-1} \end{vmatrix}$
$A_{21}(n, n+1)$	$\begin{vmatrix} 0 & 0 & \cdots & 0 & 0 & 1 \\ 0 & 0 & \cdots & 0 & 1 & 0 \\ \cdot & & \cdots & & & \cdot \\ 0 & 0 & \cdots & 0 & 0 & 0 \\ 0 & 1 & \cdots & 0 & 0 & 0 \end{vmatrix}$	$\begin{vmatrix} 0 & 1 & \cdots & 0 & 0 & 0 \\ 0 & 0 & \cdots & 0 & 0 & 0 \\ \cdot & & \cdots & & & \cdot \\ 0 & 0 & \cdots & 0 & 1 & 0 \\ 0 & 0 & \cdots & 0 & 0 & 1 \end{vmatrix}$
$A_{22}(n, n-1)$		$\begin{vmatrix} 1 & 0 & \cdots & 0 & 0 & 0 \\ -1 & 1 & \cdots & 0 & 0 & 0 \\ \cdot & & \cdots & & \cdot & \cdot \\ 0 & 0 & \cdots & 0 & -1 & 1 \\ 0 & 0 & \cdots & 0 & 0 & -1 \end{vmatrix}$

(15) Calculate the following parameters (F_j, y_j, $\bar{h}_j$, and $\bar{t}_j$) of each solution entering each evaporator (see Figs 3 and 4) by using eqns (49–51) and eqn (58) for $j = 1, 2, \ldots, (n-1)$.

As soon as the liquid stream F_j is forced through the film-forming device of the j-th evaporator, flashing may occur provided that $\bar{t}_j$ is greater than the equilibrium boiling temperature (t_{Bj}) at the prevailing pressure P_j in the same evaporator (for the sake of simplicity the pressure drop through each heat exchanger was assumed to be negligible). Of course, for $\bar{t}_j < t_{Bj}$ no flashing occurs and the calculation procedure continues from step 17.

(16) Estimate the inlet temperature (t'_j) of the liquid stream after flashing according to the following iterative process. A first estimate of $t'_j = \bar{t}_j$ is made to calculate the composition

$$y'_j = \frac{\lambda_w + c_{wv}t'_j - h'_j}{\lambda_w + c_{wv}t'_j - \bar{h}_j} y_j \tag{69}$$

and the boiling point rise $\Delta T'_{b,j} = f(t'_j, y'_j)$ of the flashed liquor, then a new estimate of t'_j is calculated by summing up the aforementioned boiling point rise to the saturation temperature of solvent at P_j and resubstituted into eqn (69), thus repeating the procedure until the difference between two subsequent estimations of t'_j is smaller than $0.5°C$.

(17) Evaluate the design heat transfer coefficient (U_{Dj}) by combining the clean HTC(U_{Cj}) and the fouling factor (R_{dj}) to take into account the additional resistance to heat transfer due to adhesion of suspended matter or incrusting components present in the thin juice inside and outside the tubes

$$1/U_{Dj} = R_{dj} + 1/U_{Cj} \tag{70}$$

where U_{Cj} is the clean overall HTC referred to the temperature difference $(T_{j-1} - t_j)$ in each evaporator calculable by means of eqns (33), and (36–40) (see Section 3.2).

(18) Check if the difference between the old (U_{Dj}) and the new (U'_{Dj}) values of the design HTC in each evaporator is less than a prefixed tolerance (ε')

$$|U'_{Dj} - U_{Dj}| \leq \varepsilon \quad \text{for} \quad j = 1, 2, \ldots, n \tag{71}$$

If the above difference is greater than ε', the process is repeated starting from step 9, by using the method of direct substitution; otherwise, the design of the MEFFE system may be retained satisfactorily and used as an intermediate step of the optimisation procedure described in the following section.

5. OPTIMISATION PROCEDURE

The design of a MEFFE system capable of concentrating a given flow rate (S_0) of a feed liquor at temperature t_0 from an initial solute composition (x_0) to a final one (x_n) requires several design parameters (number of effects (n), size—d_o, d_i, L—and closeness—P_T—of tubes and their type of pitch), operating variables (live steam condensation temperature T_0; inlet cooling water temperature t_{wi}; and recirculation ratio E_j in each effect), and a design criterion (such as that based on identical heat transfer surfaces or identical temperature differences in all the effects according to the operation in backward or forward flow) to be defined.

Implicit constraint of this optimisation implies that the wall temperature (t_p) in each FFE must be smaller than a critical value (T_c), which depends on the thermosensitivity of the liquor to be processed and the overall duration of the concentration process. In fact, improper thermal profiles during the evaporation of fruit and vegetable juices may result in a series of thermal degradation reactions, thus involving, for instance, off-taste, off-flavour, browning and nutrient loss.

With reference to Figs 3 and 4 the maximum value of t_p is encountered in the first effect and can be simply determined by assuming that the thermal resistance of the tube wall is negligible (see Fig. 2b)

$$t_{pl} = \frac{d_o h_{Tol} T_0 + d_i h_{Tl} t_1}{d_o h_{Tol} + d_i h_{Tl}} \leq T_c \qquad (72)$$

thus showing that t_{pl} is heavily dependent upon live steam condensation temperature (T_0), heat and momentum transfer conditions in the FFE.

More specifically, a generic MEFFE system of assigned water removal capacity may be optimised with respect to the number of effects (n), as follows.

(1) Assume an initial set of operating variables (T_0, t_{wi}, E_j) and select both the type of operation in reverse or forward flow and the design strategy ($A = $ const or $\Delta T = $ const).

(2) Design the MEFFE system as shown in Section 4.

(3) Compare t_{pl} with the assumed critical value (T_c). If t_{pl} is greater than T_c, find a new set of operating variables and repeat the procedure from step 2; if the contrary, the optimisation procedure can continue.

(4) Design the main equipment (i.e. the vacuum system, liquid–vapour separators and centrifugal pumps in all the effects) and determine the utility consumption (i.e. cooling water, live steam and electricity) of the system under study, as shown in the Appendix.

(5) Evaluate the investment costs of the evaporation unit (which consists of n FFE, n centrifugal pumps each one with stand-by, n liquid–vapour separators and a vacuum system), by taking into account both the bare equipment costs and auxiliary costs (instruments, piping and valves, insulation, civil work, electrical, installation, etc.). By using Guthrie's[13] concept of total module factor (ζ_i), which represents the contribution of all direct and indirect costs in the bare process module plus the contingencies necessary to adjust for unlisted items or insufficient design definition, as well as for contractor fees, the total cost of each item of the evaporation unit can be roughly

evaluated on the basis of free on board (f.o.b.) equipment costs, thus yielding the following overall cost (C_I) of the evaporation unit

$$C_I = \sum_{1}^{n} {}_j (\zeta_E C_{Ej} + 2\zeta_P C_{Pj} + \zeta_S C_{Sj}) + C_V \qquad (73)$$

where C_{Ej}, C_{Pj} and C_{Sj} are, respectively, the f.o.b. costs of the j-th FFE, centrifugal pump and liquid–vapour separator, and C_V is the installed cost of the vacuum system; all the aforementioned costs can be calculated via the correlations reported in the Appendix.

(6) Evaluate the operating costs (C_0) of the evaporation unit by taking into account the following items: investment-related (C_{Io}), utility (C_{Uo}) and labour (C_{Lo}) costs.

The first item includes depreciation and maintenance. Whereas the latter can be expressed as a percentage of about 3% of the investment costs, the former was estimated over a 15-year period (n_D) at an interest rate (i) of 15%.

The utility costs include steam, cooling water and electricity, while the labour costs can be evaluated on a basis of one seasonal skilled worker per shift whatever the number of effects at \$13 500/year and referring to a working period ($\Delta\tau$) of the plant of 120 days per year on $n_S = 3$ shifts per day. Therefore, the overall operating costs of the evaporation unit are

$$C_0 = C_{Io} + C_{Uo} + C_{Lo} \qquad (74)$$

with

$$C_{Io} = \left\{ 0\cdot03 + \frac{i(1+i)^{n_D}}{(1+i)^{n_D} - 1} \right\} C_I / \Delta\tau \qquad (75)$$

$$C_{Uo} = C_s + C_w + C_e \qquad (76)$$

$$C_s = (V_0 + V_E)c_s \qquad (77)$$

$$C_w = ((W_1 + W_2)/\rho_w)c_w \qquad (78)$$

$$C_e = \sum_{1}^{n} {}_j N_{Pj} c_e \qquad (79)$$

$$C_{Lo} = (n_S + 1)13\,500/\Delta\tau \qquad (80)$$

where $c_s(= c\$2\cdot5/\text{kg})$, $c_w(= \$0\cdot1/\text{m}^3)$, and $c_e(= c\$5/\text{kWh})$ are, respectively, the specific costs of steam, cooling water and electric power, while the other symbols are reported in the Nomenclature section.

(7) Determine the operating conditions associated with the minimum overall operating costs (C_0) of the evaporation unit examined as

a function of the number of effects (n) and the aforementioned design and operating parameters, by repeating the procedure from step 2.

6. OPTIMAL DESIGN OF MEFFE FOR CONCENTRATED APPLE JUICE PRODUCTION

The optimisation procedure outlined in the previous section has been applied to the production of concentrated apple juices, by assuming as a basis of computation a typical plant of large size with an evaporation capacity of about 40 000 kg of depectinised, clarified apple juice per hour in the range 11·9–71°Brix.[43]

All the input data required for this study are summarised in Table 4.

More specifically, the main design parameters of the FFE system examined coincide with those of the FFE manufactured by Officine Metalmeccaniche Santoro (Messina, Italy), while a fouling factor R_d equal to 4×10^{-4} m² °C/W was chosen according to Ref. 1.

As far as the critical temperature (T_c) for concentrated apple juices is concerned, no data appear to be reported in the literature; so its rough value can be only extracted from the information available on

TABLE 4
INPUT DATA REQUIRED FOR APPLE-JUICE MEFFE DESIGN

Parameter		Value	Unit
(1) Feed flow rate	(S_0)	40 000	kg/h
(2) Feed temperature	(t_0)	30	°C
(3) Feed concentration	(x_0)	11·9	°Brix
(4) Output concentration	(x_n)	71	°Brix
(5) Live steam temperature	(T_0)	90–100	°C
(6) Cooling water temperature range	$(t_{wi} - t_{wo})$	21–33	°C
(7) Thermal loss	(β_j)	0·03	—
(8) Recirculation ratio	(E_j)	0–5	—
(9) Tube inside/outside diameter	(d_i/d_o)	39/42.5	mm
(10) Tube length	(L)	10	m
(11) Tube closeness	(P_T)	47	mm
(12) Type of pitch	—	Δ	—
(13) Fouling factor	(R_d)	4×10^{-4}	m² °C/W
(14) Operation	—	$\begin{cases} \text{co-current} \\ \text{countercurrent} \end{cases}$	—
(15) Design strategy	—	$\begin{cases} A = \text{Const} \\ \Delta T = \text{Const} \end{cases}$	—

the Maillard-type browning reaction and ascorbic acid oxidation in fruit juices.

In apple juice the former reaction, being characterised by an activation energy of 113 kJ/mole and a first-order reaction rate constant of 8.5×10^{-3}/min at 121°C,[44] would involve a 99% retention of the quality factor concerned on condition that the mean residence time in the evaporator is shorter than 22·5 min at 90°C or 8·2 min at 100°C.

The latter reaction, being greatly sensitive to air or oxygen, light and heat in presence of traces of copper and iron,[45] might result in 25–50% of vitamin loss owing to improper juice deaeration prior to concentration, thus making it sometimes compulsory to fortify the commercial processed juice.[46]

However, under the time-temperature conditions given above, the available data on ascorbic acid oxidation (activation energies of 70–110 kJ/mole and first-order reaction rate constants of $2–4 \times 10^{-7}$/ min at 20°C[12,47]) involve as a consequence no less than 95% ascorbic acid retention, thus showing that the most perceptible change in product quality is due to ascorbic acid oxidation.

Therefore, owing to liquor residence times in the recirculation MEFFE system shorter than 10 min,[11,12] no deleterious effect on product quality is expected when operating at temperatures below 90°C.

A synopsis of the regression equations used to predict the physical properties of depectinised, clarified apple juices necessary for this study (e.g. boiling point rise, ΔT_b, rheological behaviour; density, ρ; specific heat, c_p; thermal conductivity, k; and surface tension, σ) is shown in Table 5. In particular, k and σ, being unknown for the juice examined, were assumed to be equal to those of aqueous sucrose solutions: the k and σ values were therefore extracted from Honig[48] and correlated by means of the method of least squares with mean standard errors smaller than 1·5% (see Table 5).

Finally, owing to the present upward surge of the US currency, the cost correlations shown in the Appendix were not updated by introducing the actual value of the *Chemical Engineering Plant Cost Index* in order to allow our costings to be consistent with the present Italian market prices for the main items of the evaporation units examined. Thus, in this way the overall operating costs of the evaporation units under study were found to be in line with those recently presented by Renshaw *et al.*[60]

TABLE 5
REGRESSION EQUATIONS OF THE PARAMETERS NECESSARY TO PREDICT THE
ENGINEERING PROPERTIES OF CONCENTRATED APPLE JUICES FOR WEIGHT
FRACTIONS OF EQUIVALENT SUCROSE (x) RANGING FROM 0 TO 0·748 (AS
EXTRACTED FROM MORESI AND SPINOSI[40])

Physical property	Regression	Unit
Boiling point rise	$\Delta T_b = (A - BT_w)T_w/(B_0 + BT_w)$	K
	$A = 14\cdot85x$	K
	$B_0 = -5234$	K
	$B = 0\cdot383x - 1\cdot122x^2 + 1\cdot871x^3$	—
Rheological behaviour	$\tau = K\gamma^v$	Pa
	$K = \exp(C + D/T_K)$	Pa s^v
	$C = -13\cdot47 + 4\cdot88X - 2\cdot38X^2$	—
	$D = 1926\cdot48 + 588X - 873X^2 +$	K
	$\quad 700X^3 - 106\cdot7X^4$	
	$v = 1 - 0\cdot186X + 0\cdot047X^2$	—
	$X = x/(1 - x)$	—
Specific heat	$c_p = xc_{pa} + (1 - x)c_{wl}$	J/(g K)
	$c_{pa} = 1\cdot1$	J/(g K)
	$c_{wl} = 4\cdot186$	J/(g K)
Density	$\rho = 1/\{x/\rho_a + (1 - x)/\rho_w\}$	kg/m^3
	$\rho_a = 1619$	kg/m^3
	$\rho_w = 998$	kg/m^3
Thermal conductivity	$k^a = 0\cdot213 + 1\cdot316 \times 10^{-3}T - 0\cdot339x$	W/(m K)
Surface tension	$\sigma^a = 0\cdot0727 + 1\cdot1 \times 10^{-4}x$	N/m

^{a} Referred to aqueous sucrose solutions and estimated by fitting Honig's
data.[48]

6.1. Optimal Number of Effects
In order to minimise the overall operating costs of apple juice
concentration, several calculations based on the aforementioned
design strategies were carried out by varying the number of effects (n)
or the method of feeding of the evaporation unit.

Tables 6 and 7 show the main responses of the mathematical model
previously described, that is primary steam (V_0 and V_E), cooling water
($W_1 + W_2$) and electric power ($\sum_j N_{Pj}$) consumption, the overall heat
transfer surface ($\sum_j A_j$), the f.o.b. costs of all FFE ($\sum_j C_{Ej}$), liquid–
vapour separators ($\sum_j C_{Sj}$) and centrifugal pumps ($\sum_j C_{Pj}$), the installed
costs of the vacuum system (C_V), the total investment costs (C_I) and
finally the overall operating costs (C_0), for a series of evaporation units
consisting of two to nine effects, operating either co-currently or
countercurrently and designed by assuming either identical heat

TABLE 6

PRODUCTION OF CONCENTRATED APPLE JUICES BY MEANS OF SEVERAL EVAPORATION UNITS OPERATING IN FORWARD FLOW WITH $T_0 = 95°C$, $t_0 = 30°C$ AND $E_j = 0$ FOR $j = 1, 2, \ldots, n$ AND DESIGNED ON THE BASIS OF IDENTICAL HEAT TRANSFER SURFACES (a) OR IDENTICAL TEMPERATURE DIFFERENCE (b) IN ALL THE EFFECTS: MAIN DESIGN AND OPERATING PARAMETERS us THE NUMBER OF EFFECTS (n)

Parameter		Number of effects (n)								Unit
		2	3	4	5	6	7	8	9	
Utility consumption										
V_0	a.	20631	15091	12321	10653	9542	8748	8154	7694	kg/h
	b.	20393	14839	12609	10411	9306	8518	7928	7470	kg/h
V_E	a.	52	76	100	123	147	171	195	219	kg/h
	b.								212	kg/h
$W_1 + W_2$	a.	902	603	455	366	307	265	234	210	m³/h
	b.	899	602	455	367	308	267	236	219	m³/h
$\sum_j N_{Pj}$	a.	9	13	17	20	24	28	32	36	kW
	b.									
OHTS $\sum_j A_j$	a.	1028	1498	1978	2453	2944	3449	3978	4533	m²
	b.	1045	1532	1991	2465	2954	3458	3988	4539	m²
Investment costs										
$\sum_j C_{Ej}$	a.	233	339	448	555	666	781	901	1027	10³ $
	b.	238	350	455	564	677	793	916	1043	10³ $
$\sum_j C_{Sj}$	a.	149	142	143	149	155	162	169	177	10³ $
	b.	161	153	155	160	166	173	180	187	10³ $
$\sum_j C_{Pj}$	a.	9	13	17	21	26	30	34	38	10³ $
	b.									
C_V	a.	32	36	40	43	46	48	50	52	10³ $
	b.									
C_I	a.	1397	1667	1981	2305	2645	2995	3362	3745	10³ $
	b.	1554	1924	2309	2720	3148	3590	4055	4536	10³ $
Operating costs C_0	a.	737·7	591·6	533·2	509·3	503·5	508·1	520·1	537·5	$/h
	b.	743·9	605·6	553·1	536·3	537·7	549·9	569·8	595·0	$/h

OHTS = overall heat transfer surface.

TABLE 7

PRODUCTION OF CONCENTRATED APPLE JUICES BY MEANS OF SEVERAL EVAPORATION UNITS OPERATING IN BACKWARD FLOW WITH $T_0 = 90°C$, $t_0 = 30°C$, AND $E_j = 0$ FOR $j = 1, 2, \ldots, n$ AND DESIGNED ON THE BASIS OF IDENTICAL HEAT TRANSFER SURFACES (a) OR IDENTICAL TEMPERATURE DIFFERENCE (b) IN ALL THE EFFECTS: MAIN DESIGN AND OPERATING PARAMETERS us THE NUMBER OF EFFECTS (n)

Parameter		2	3	4	5	6	7	8	9	Unit
Utility Consumption										
V_0	a.	19004	13145	10208	8446	7274	6440	5818	5338	kg/h
	b.	19038	13193	10263	8505	7335	6503	5882	5402	
V_E	a.	54	79	104	128	153	178	203	227	kg/h
	b.									
$W_1 + W_2$	a.	867	551	392	297	234	189	156	130	m³/h
	b.	866	549	389	292	228	182	147	121	
$\sum_j N_{Pj}$	a.	8	12	16	20	24	28	32	36	kW
	b.									
OHTS $\sum_j A_j$	a.	1053	1541	2053	2585	3139	3716	4315	4933	m²
	b.	1042	1524	2028	2551	3095	3647	4215	4807	
Investment costs										
$\sum_j C_{Ej}$	a.	239	349	465	586	712	843	980	1121	10³ $
	b.	236	346	460	580	704	831	962	1099	
$\sum_j C_{Sj}$	a.	186	177	178	183	189	196	203	211	10³ $
	b.	177	169	170	174	180	186	193	200	
$\sum_j C_{Pj}$	a.	9	13	17	21	26	30	34	39	10³ $
	b.									
C_V	a.	32	37	40	43	46	49	51	53	10³ $
	b.									
C_I	a.	1572	1847	2176	2533	2910	3305	3714	4136	10³ $
	b.	1618	1976	2390	2833	3299	3775	4266	4778	
Operating costs C_0	a.	707·4	552·1	489·8	465·5	460·7	467·6	482·1	501·9	$/h
	b.	711·9	563·4	507·9	490·3	492·5	505·8	526·7	553·6	

transfer surface (a) or identical temperature difference (b) in all the effects.

Whatever the method of feeding, the minimum value of C_0 is associated with the operation of a six-effect FFE system designed on the basis of the former strategy ($A = $ const). In contrast, the latter strategy ($\Delta T = $ const), which involves a specific project for each FFE, results in larger fixed-capital requirements owing to the contribution of the engineering costs, which have to be considered for all the evaporators and not for a single FFE as is done when the strategy of identical heat transfer surfaces is applied. For this reason, the minimum value of manufacturing costs for such a strategy corresponds to the operation of a five-effect FFE system. In other words, this does mean nothing else but that the optimal number of effects derives from the balance between decreasing operating costs and increasing capital charges.

Since the strategy of identical heat transfer surfaces yields smaller investment and operating costs whatever the number of effects, it will be further considered to assess the optimal operation of the evaporation unit under study.

As far as the method of feeding is concerned, backward feeding has the advantage of smaller operating costs, owing to its improved steam economy.

In fact, since the more viscous material meets increasingly hotter surfaces in going from effect to effect, the increase in viscosity on concentration is partially off-set by the higher temperatures experienced, thus resulting in higher heat transfer coefficients. This can however bring about localised overheating, unless the wall temperature is kept lower than a critical value (see Section 5) by throttling the live steam valve at appropriate final pressures (Table 8).

In contrast to classic co-current multiple-effect evaporation units, that have the advantage of not requiring interstage pumps over the countercurrent counterparts, either co-current or countercurrent MEFFE systems need interstage pumps to raise the liquor from the bottom of a generic liquid–vapour separator to the top of the near FFE and overcome pipe and fittings friction. In this specific case, the pressures at two adjacent effects have no practically significant influence on interstage pump ratings, nor on pump electric power consumption.

Although with forward flow the design heat transfer coefficient decreases during passage through the plant (see for instance Table 8)

TABLE 8

OPTIMAL OPERATING CONDITIONS FOR THE PRODUCTION OF CONCENTRATED APPLE JUICES BY USING A SINGLE-PASS SIX-EFFECT FFE SYSTEM OPERATING IN FORWARD (a) OR BACKWARD (b) FLOW

Parameter		1st effect	2nd effect	3rd effect	4th effect	5th effect	6th effect	Unit
Internal pressure, P_j	a.	57·5	45·8	35·4	26·1	17·1	6·0	kPa
	b.	34·1	22·4	15·5	11·0	7·9	5·8	
Juice temperature, t_j	a.	85·7	80·1	74·0	67·0	58·4	42·7	°C
	b.	80·2	64·0	55·7	48·5	42·0	36·1	
Boiling point rise, ΔT_{bj}	a.	0·75	0·81	0·88	1·01	1·49	6·4	°C
	b.	8·07	1·35	0·87	0·71	0·61	0·53	
Max. wall temperature, t_{pj}	a.	89·2	82·8	77·2	71·1	64·4	55·9	°C
	b.	88·1	70·1	60·3	52·4	45·5	39·2	
Input liq. flow rate, S_{j-1}	a.	40	34·64	29·15	23·57	17·95	12·33	Mg/h
	b.	13·47	19·77	25·58	30·9	35·72	40	
Input liq conc., x_{j-1}	a.	11·9	13·7	16·3	20·2	26·5	38·6	°Brix
	b.	35·3	24·1	18·6	15·4	13·3	11·9	
Output liq. flow rate, S_j	a.	34·64	29·15	23·57	17·95	12·33	6·70	Mg/h
	b.	6·7	13·47	19·77	25·58	30·9	35·72	
Output liq. conc., x_j	a.	13·7	16·3	20·2	26·5	38·6	71	°Brix
	b.	71	35·3	24·1	18·6	15·4	13·3	
Design heat transfer coeff., U_{Dj}	a.	1540	1333	1248	1122	901	492	W/(m²K)
	b.	874	992	1087	1121	1132	1135	
Heat transfer surface, A_j	a.			513				m²
	b.			523				
Live steam cond. temp., T_0	a.			93				°C
	b.			90				
C_0	a.			508·6				$/h
	b.			460·7				

owing to both an increase in concentration and decrease in temperature in each effect, thermosensitive liquors undergoing co-current
evaporation tend to be more efficiently preserved from heat damage,
especially because the wall temperature in the effects operating at
lower pressures reduces as juice concentrates. In actual fact, there is
limited, if any, utilisation of backward flow practice in the food
industry.

However, since over the last decades HTST processing techniques,
yielding better product quality retention with respect to conventional
processing, have gained wide acceptance, it is not inconceivable that
countercurrent evaporation performed under strict wall temperature
control might result in smaller manufacturing costs without affecting
quality.

In fact, the conditions in the later stages of evaporation do not seem
to matter as far as protein denaturation (which usually happens at a
temperature increasing with concentration) is concerned, though these
are important when considering other factors such as browning. In
such cases, it is possible to rely on shorter residence times only to keep
flavour, colour, texture and vitamin content minimally altered.

The optimal operating conditions and design parameters for the
plant size examined are shown in Table 8 and refer to a six-effect FFE
system operating in forward (a) or backward (b) flow.

It can be noted that the live steam condensation temperature in the
co-current evaporation unit was reduced to 93°C to maintain the
maximum wall temperature (t_{pl}) within the constraint of T_c. Moreover,
by comparing either the design heat transfer coefficients or the overall
heat transfer surface in forward and backward flow, it is possible to
stress that the smaller operating costs of the countercurrent six-effect
FFE system exclusively derive from better steam economy rather than
smaller fixed-capital requirements.

6.2. Feed Pre-heating

To verify the stability of the above values of C_0 with respect to the
input temperature (t_0) of raw juice, a sensitivity analysis was carried
out by varying t_0 stepwise. Of course, any value of t_0 greater than 30°C
would involve a pre-heating operation to be carried out before feeding
the evaporation unit, thus charging the evaporation unit with extra
costs (i.e. the live steam consumed to preheat the feed liquor and the
investment-related costs for the preheater itself).

In our specific case, such an operation might be safely carried out by means of a plate-and-frame heat exchanger, where live steam condensing at 93°C is used as hot fluid so as to maintain the wall temperature at values smaller than T_c.

By assuming a design overall heat transfer coefficient of about $2 \, \text{kW}/(\text{m}^2 \, \text{K})$,[61] it was possible to estimate the heat transfer surface (A_{PHE}) with a 10% overdesign factor and then evaluate the f.o.b. cost (C_{PHE}) of the plate heat exchanger (PHE) via the following correlation[62]

$$C_{PHE} = 634 \cdot 5 A_{PHE}^{0 \cdot 778} \qquad (81)$$

The installed cost of PHE was finally estimated referring to a total module factor of 2·5.

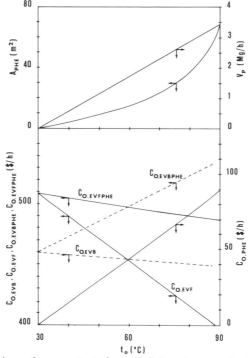

FIG. 5. Production of concentrated apple juices in two single-pass six-effect FFE units, respectively, operating in forward and backward flow, fed by raw juice preheated up to t_0 ranging from 30° to 90°C: heat transfer surface (A_{PHE}) of the plate heat exchanger used for the pre-heating operation; flow rate (V_p) of live steam condensing at 93°C; operating costs of the pre-heating section $(C_0)_{PHE}$, co-current evaporation unit $(C_0)_{EVF}$, countercurrent evaporation unit $(C_0)_{EVB}$, and combined units $(C_0)_{EVFPHE}$ and $(C_0)_{EVBPHE}$ vs inlet feed temperature (t_0).

Figure 5 shows the effect of t_0 on the following parameters: live steam consumed for liquor pre-heating, A_{PHE}, operating costs of the pre-heating section $(C_0)_{PHE}$ and six-effect FFE unit operating either in forward $(C_0)_{EVF}$ or in backward $(C_0)_{EVB}$ flow, and overall operating costs of the two combined units, that is, $(C_0)_{EVFPHE}$ and $(C_0)_{EVBPHE}$.

It is worth pointing out that feed pre-heating is advantageous in the co-current MEFFE units and disadvantageous in the countercurrent MEFFE ones. In fact, in backward feeding raw juice pre-heating just results in more flashing in the lower pressure stage with the overall effect of overcharging the barometric condenser and simultaneously lowering the re-use of that steam with respect to the case of no feed-pre-heating.

To clarify such a statement, Tables 9 and 10 report the main design

TABLE 9

MAIN DESIGN AND OPERATING PARAMETERS OF A SINGLE-PASS SIX-EFFECT FFE SYSTEM DESIGNED ON THE BASIS OF IDENTICAL HEAT TRANSFER SURFACES AND OPERATING IN FORWARD OR BACKWARD FLOW WITH APPROPRIATE FEED PRE-HEATING

Parameter	Backward flow	Forward flow	Unit
T_0	90	93	°C
t_0	80	85	°C
Utility consumption			
V_P	2869	3155	kg/h
V_0	6659	5731	kg/h
V_E	153	147	kg/h
$W_1 + W_2$	370	308	m³/h
$\sum_j N_{Pj}$	23	24	kW
Heat transfer surface			
A_{PHE}	37	48	m²
$\sum_j A_j$	2931	2906	m²
Investment costs			
C_{PHE}	10·5	12·9	10^3 $
$\sum_j C_{Ej}$	663	657	10^3 $
$\sum_j C_{Sj}$	194	156	10^3 $
$\sum_j C_{Pj}$	26	26	10^3 $
C_V	46	46	10^3 $
$(C_I)_{EV}$	2802	2627	10^3 $
Operating costs			
$(C_0)_{EV}$	450·2	406·8	$/h
$(C_0)_{PHE}$	73·8	81·4	$/h
$(C_0)_{EVPHE}$	524·0	488·2	$/h

TABLE 10

OPERATING CONDITIONS FOR THE PRODUCTION OF CONCENTRATED APPLE JUICES IN A SINGLE-PASS SIX-EFFECT FFE SYSTEM OPERATING IN FORWARD (a) OR BACKWARD (b) FLOW WITH APPROPRIATE FEED PRE-HEATING

Parameter		1st effect	2nd effect	3rd effect	4th effect	5th effect	6th effect	Unit
Internal pressure, P_j	a.	62·5	49·5	38	27·6	17·9	6·0	kPa
	b.	34·1	22·4	15·5	10·9	7·9	5·8	
Juice temperature, T_j	a.	87·8	82	75·6	68·4	59·3	42·7	°C
	b.	80·3	64·1	55·7	48·5	42	36·1	
Boiling point rise, ΔT_{bj}	a.	0·76	0·82	0·89	1·02	1·49	6·4	°C
	b.	0·55	0·63	0·74	0·91	1·44	8·07	
Max wall temp., T_{pj}	a.	90·5	84·8	78·9	72·6	65·6	56·7	°C
	b.	88·1	70·2	60·4	52·5	45·5	39·3	
Input liq. flow rate, S_{j-1}	a.	40	34·67	29·19	23·62	17·99	12·35	Mg/h
	b.	12·89	18·64	23·95	28·8	33·2	40	
Input liq. conc., x_{j-1}	a.	11·9	13·7	16·3	20·2	26·5	38·5	°Brix
	b.	36·9	25·5	19·9	16·5	14·3	11·9	
Output liq. flow rate, S_j	a.	34·67	29·19	23·62	17·99	12·35	6·7	Mg/h
	b.	6·7	12·89	18·64	23·95	28·80	33·2	
Output liq. conc., x_j	a.	13·73	16·3	20·2	26·5	38·5	71	°Brix
	b.	71	36·9	25·5	19·9	16·5	14·3	
Output vap. flow rate, V_j	a.	5335	5473	5571	5627	5644	5646	kg/h
	b.	6185	5755	5309	4852	4393	6802	
Design heat transfer coeff., U_{Dj}	a.	1398	1341	1257	1131	907	490	W/(m² K)
	b.	862	971	1069	1106	1119	1106	

and operating parameters, as well as the main processing variables, of a six-effect FFE unit (designed on the basis of identical heat transfer surface in all the effects) operating in forward or backward flow with feed liquor pre-heated at 85°C and 80°C, respectively.

6.3. Liquor Recirculation

In using multiple-effect evaporators the quantity of vapour that is re-used is almost the same in each effect (see Table 10); so, if a high concentration ratio is used, in the effects treating the progressively more concentrated liquor there may not be enough liquid to wet the tube wall, thus giving rise to the need for recirculation.

Unfortunately, such an operation cannot be regarded as ineffectual on product quality and water removal costs. In fact, it *a priori* gives rise to longer residence times of liquor in the evaporation unit and larger investment and operating costs, essentially because with recirculation FFE interstage pumps have to be oversized with respect to single-pass FEE.

To define the optimal value of the recirculation ratio E_l in the generic l-th effect of an MEFFE system, specific experimental information about liquor spreadability over FFE tubes and product deterioration is prerequisite for the mathematical model under examination.

By assuming for the sake of simplicity that in each effect constant input flow rate is maintained and the same amount of solvent is vaporised, the recirculation ratio E_l can be estimated as follows

$$E_l = \frac{(l-1)(1-x_0/x_n)}{n-l(1-x_0/x_n)} \quad \text{for} \quad l = 1, 2, \ldots, n \quad (82)$$

where l is the dummy index expressed by eqn (44) or eqn (45) provided that the MEFFE system examined is operating in forward or backward flow.

Under the above assumption and among the recirculation MEFFE systems capable of fulfilling the specifications of Table 4, a recirculation six-effect FFE unit designed on the basis of identical heat transfer surface was found to yield minimum manufacturing costs, though greater than those associated with single-pass evaporation.

Moreover, similarly to previous single-pass MEFFE systems, the operation of the countercurrent unit was more convenient than that of the co-current one. However, the difference between the two values of C_0 can be minimised by feeding the co-current unit with raw juice at almost its boiling point at the pressure existing in the first effect.

TABLE 11
PRODUCTION OF CONCENTRATED APPLE JUICES
IN A RECIRCULATION SIX-EFFECT FFE SYSTEM
OPERATING IN BACKWARD OR FORWARD FLOW
WITH APPROPRIATE FEED PRE-HEATING

Parameter	Backward flow	Forward flow	Unit
T_0	90	92	°C
t_0	30	85	°C
Utility consumption			
V_P	—	3155	kg/h
V_0	7184	5849	kg/h
V_E	153	147	kg/h
$W_1 + W_2$	238	306	m³/h
$\sum_j N_{Pj}$	38	40	kW
Heat transfer surface			
A_{PHE}	—	48	m²
$\sum_j A_j$	3689	3726	m²
Investment costs			
C_{PHE}	—	12·9	10^3 $
$\sum_j C_{Ej}$	843	852	10^3 $
$\sum_j C_{Sj}$	194	151	10^3 $
$\sum_j C_{Pj}$	26	26	10^3 $
C_V	46	46	10^3 $
$(C_I)_{EV}$	3285	3129	10^3 $
Operating costs			
$(C_0)_{EV}$	489·4	450·3	$/h
$(C_0)_{PHE}$	—	81·5	$/h
$(C_0)_{EVPHE}$	489·4	531·8	$/h

Tables 11 and 12 list the optimal values of the main processing variables and design parameters for both types of flow arrangement of a recirculation six-effect FFE unit fed with appropriately pre-heated raw juice.

7. CONCLUSIONS

Following a brief review of the heat and momentum transfer capabilities of FFE, a design strategy for MEFFE units operating in forward or backward flow was established so as to minimise water removal costs without affecting final product quality.

In particular, the effects of temperature and time upon the

TABLE 12

OPERATING CONDITIONS FOR THE PRODUCTION OF CONCENTRATED APPLE JUICES IN A RECIRCULATION SIX-EFFECT FFE SYSTEM OPERATING IN FORWARD (a) OR BACKWARD (b) FLOW WITH APPROPRIATE FEED PRE-HEATING

Parameter		1st effect	2nd effect	3rd effect	4th effect	5th effect	6th effect	Unit
Internal pressure, P_j	a.	62·6	51·8	42	32·9	23·7	6·0	kPa
	b.	27·7	19	13·7	10·1	7·6	5·8	
Juice temperature, T_j	a.	87·9	83·2	78·1	72·4	65·5	42·7	°C
	b.	75·2	60·4	53·1	46·9	41·3	36·1	
Boiling point rise, ΔT_{bj}	a.	0·76	0·82	0·91	1·05	1·56	6·4	°C
	b.	7·84	1·33	0·86	0·70	0·60	0·53	
Max wall temp., T_{pj}	a.	90	85·4	80·6	75·5	69·9	63·2	°C
	b.	89	65·5	57	50·3	44·3	38·9	
Input liq. flow rate, S_{j-1}	a.	40	34·55	29·02	23·45	17·87	12·32	Mg/h
	b.	13·37	19·61	25·41	30·75	35·65	40	
Input liq. conc., x_{j-1}	a.	11·9	13·8	16·4	20·3	26·6	38·6	°Brix
	b.	35·6	24·3	18·7	15·5	13·4	11·9	
Output liq. flow rate, S_j	a.	34·56	29·04	23·47	17·89	12·33	6·7	Mg/h
	b.	6·7	13·37	19·61	25·41	30·75	35·65	
Output liq. conc., x_j	a.	13·8	16·4	20·3	26·6	38·6	71	°Brix
	b.	71	35·6	24·3	18·7	15·5	13·4	
Output vap. flow rate, V_j	a.	5450	5528	5571	5577	5552	5618	kg/h
	b.	6663	6240	5800	5347	4891	4355	
Recirculation ratio, E_j	a.	0	0·19	0·48	0·94	1·81	4·14	kg/kg
	b.	4·14	1·81	0·94	0·48	0·19	0	
Design heat transfer coeff., U_{Dj}	a.	1406	1358	1289	1178	958	266	W/(m²K)
	b.	486	972	1089	1124	1133	1136	

organoleptic and nutritional properties of the juice undergoing evaporation were empirically described by means of a single lumped-
parameter (i.e. juice critical temperature T_c), thus enabling the upper
limit of tube wall temperature in each FFE to be set up.

With specific reference to large scale production of concentrated
apple juices, the minimum value of the overall operating costs was
associated with the operation of a six-effect FFE unit designed on the
basis of identical heat transfer surfaces (to save engineering costs) and
operating in backward flow (to rely on better steam economy).

Since in the food industry there is practically no utilisation of
backward feeding, the above result, which implicitly derives from the
constraint established by the aforementioned lumped-parameter,
might be regarded as the outcome of an oversimplification of the
complex set of physico-chemical and biochemical phenomena responsible for thermal damage of concentrated fruit juices.

Nevertheless, it is not inconceivable that shorter residence times of
concentrated juice at temperatures higher than those experienced in
classic cocurrent evaporation units, but still lower than T_c, will lead to
a very much more economic application and, what is more, less
product deterioration.

When experimental evidence proves that backward feeding is not
suitable for the product examined, the more traditional MEFFE unit
with forward feeding tends to be the only choice.

In such a case, feed pre-heating almost up to the liquor boiling point
at the pressure maintained into the first effect seems to be the only
way to narrow the gap in the operating costs of the two aforementioned types of flow arrangement. More specifically, it was found that
under optimal conditions that difference represents as little as 9% of
the overall operating costs (i.e. 1·47 cents per kg of water removed) of
the more economic six-effect FFE unit, thus making any industrial
practice aimed at overemphasising economy at the expense of quality
definitely short-sighted.

In conclusion, it is worthwhile pointing out that the mathematical
model here described can be used not only to design and optimise a
new evaporation unit, but also to calculate the suitability of an existing
evaporation unit for new process conditions and choose the main
process variables (i.e. input feed temperature, recirculation ratios,
number of effects, live steam condensation temperature, etc.) so as to
minimise the overall manufacturing costs without disregarding product
quality.

REFERENCES

1. ANGELETTI, S. and MORESI, M., *J. Food Technol.*, 1983, **18**, 539–63.
2. NARAYANA MURTHY, V. and SARMA, P. K., *Can. J. Chem. Eng.*, 1977, **55**, 732–5.
3. HEID, J. L. and CASTEN, J. W. In *Fruit and Vegetable Juice Processing Technology*, 1st edn, 1961, ed. by D. K. Tressler and M. A. Joslyn, AVI Publishing Co., Westport.
4. BERRY, R. E. and VELDHUIS, M. K. In *Citrus Science and Technology*, Vol. 2, 1977, ed. by S. Nagy, P. E. Shaw and M. K. Veldhuis, AVI Publishing Co., Westport.
5. VARSEL, C. In *Citrus Nutrition and Quality*, 1980, ed. by S. Nagy and J. A. Attaway, *ACS Symp. Series No. 143*, American Chemical Society, Washington D.C.
6. SANTORO, A. (Officine Metalmeccaniche Santoro SpA, Messina, Italy), personal communication, 1984.
7. CARTER, A. L. and KRAYBILL, R. R., *Chem. Eng. Prog.*, 1966, **62**, 99–110.
8. PURCARO, F. P., *Quad. ingegn. chim. ital.*, 1979, **15**, 52–65.
9. MORGAN, A. I., JR., *Food Technol.*, 1967, **21**, 1353–7.
10. LUND, D. and SANDU, C., In *Proceedings of Fundamentals and Applications of Surface Phenomena Associated with Fouling/Cleaning in Food Processing*, 1981, ed. by B. Hallström, D. Lund and C. Tragärdh, Reprocentralen, Lund.
11. MANNHEIM, C. H. and PASSY, N., In *Advances in Preconcentration and Dehydration of Foods*, 1974, ed. by A. Spicer, Applied Science Publishers, London.
12. THIJSSEN, H. A. C. and VAN OYEN, N. S. M., In *7th European Symposium—Food: Product and Process Selection in the Food Industry. Preprints.* Eindhoven, The Netherlands, 21–23 September, 1977, European Federation of Chemical Engineering and IUFoST.
13. GUTHRIE, K. M., *Chem. Eng.*, 1969, **76**, 114–42.
14. OOSTHUIZEN, P. H. and CHEUNG, T., *J. Heat Transfer, Trans. ASME*, Series C, 1977, **99**, 152–5.
15. WHITAKER, S. and CERRO, R. L., *Chemical Eng. Sci.*, 1974, **29**, 963–5.
16. PIERSON, F. W. and WHITAKER, S., *Ind. Eng. Chem. Fundam.*, 1977, **16**, 401–8.
17. SKELLAND, A. H. P., *Non-Newtonian Flow and Heat Transfer*, 1967, John Wiley & Sons, New York.
18. FULFORD, G. D., In *Advances in Chemical Engineering*, Vol. 5, 1964, ed. by T. B. Drew, J. W. Hoopes Jr and T. Vermeulen, Academic Press, New York.
19. CHUN, K. R. and SEBAN, R. A., *J. Heat Transfer, Trans. ASME*, Series C, 1971, **93**, 391–6.
20. CHUN, K. R. and SEBAN, R. A., *J. Heat Transfer, Trans. ASME*, Series C, 1972, **94**, 432–6.

21. DUKLER, A. E. and BERGELIN, O. P., *Chem. Eng. Prog.*, 1952, **48,** 557–63.
22. SARMA, P. K., JUBOURI ALI-S., A. P. and NARAYANA MURTHY, V., *Can. J. Chem. Eng.,* 1978, **56,** 639–42.
23. DUKLER, A. E., *Chem. Eng. Prog. Symp. Ser.*, 1960, **56,** 1–10.
24. NARAYANA MURTHY, V. and SARMA, P. K. *J. Chem. Eng. Jap.*, 1973, **6,** 457–9.
25. MILLS, A. F. and CHUNG, D. K., *Int. J. Heat Mass Transfer, Trans. ASME*, Series C, 1973, **98,** 315–8.
26. BRUMFIELD, L. K. and THEOFANOUS, T. G., *J. Heat Transfer, Trans. ASME*, Series C, 1976, **98,** 496–502.
27. SEBAN, R. A. and FAGHRI, A., *J. Heat Transfer, Trans. ASME*, Series C, 1976, **98,** 315–8.
28. MCADAMS, W. H., *Heat Transmission*, 3rd ed., 1954, McGraw-Hill, London.
29. GARWIN, L. and KELLY, E. W., *Ind. Eng. Chem. Fundam.*, 1955, **47,** 392–5.
30. WILKE, W., *Ver. Deut. Ingr., Forschungsheft,* 1962, **490.**
31. AHMED, S. Y. and KAPARTHI, R., *Ind. J. Technol.*, 1963, **1,** 377–81.
32. HERBERT, L. S. and STERNS, U. J., *Can. J. Chem. Eng.*, 1968, **46,** 401–7, 408–12.
33. VON KÁRMÁN, T., *Trans. ASME.*, 1939, **61,** 705–10.
34. BELKIN, H. H., MACLEOD, A. A., MONRAD, C. C. and ROTHFUS, R. R., *AIChE. J.*, 1959, **5,** 245–8.
35. RAO, M. A., *J. Text. Stud.*, 1977, **8,** 135–68.
36. KERN, D. O., *Process Heat Transfer*, 1950, McGraw-Hill Book Co., New York.
37. HOLLAND, C. D., *Fundamentals and Modelling of Separation Processes,* 1975, Prentice-Hall, Englewood Cliffs.
38. VARSHNEY, N. N. and BARHATE, V. D., *J. Food Technol.*, 1978, **13,** 225–33.
39. MORESI, M. and SPINOSI, M., *J. Food Technol.*, 1980, **15,** 265–76.
40. MORESI, M. and SPINOSI, M., In *Engineering and Food. Vol. 1— Engineering Sciences in the Food Industry*, 1984, ed. by B. M. McKenna, Elsevier Applied Science Publishers, London.
41. MORESI, M. and SPINOSI, M., *J. Food Technol.* 1984, **19,** 519–33.
42. LAPIDUS, L., *Digital Computation for Chemical Engineers*, 1962, McGraw-Hill Book Co., New York.
43. ANON., *Food Eng. Int'l.*, 1982, **July/August,** 50–2.
44. LUND, D., In *Principles of Food Science, Part II Physical Principles of Food Preservation*, 1975, ed. by O. R. Fennema, Marcel Dekker Inc., New York.
45. HARRIS, R. S. and VON LOESECKE, H., *Nutritional Evaluation of Food Processing*, 1960, John Wiley & Sons Inc., New York.
46. AITKEN, H. C., In *Fruit and Vegetable Juice. Processing Technology*, 1961, ed. by D. K. Tressler and M. A. Joslyn, AVI Publishing Co., Westport.
47. THOMPSON, D. R., *Food Technol.*, 1982, **36,** 97–108.

48. HONIG, P., *Principles of Sugar Technology*. Vol. 1, 1953, Elsevier Publishing Co., Amsterdam.
49. PUROHIT, G. P., *Chem. Eng.*, 1983, **90**, 56–67.
50. CORRIPIO, A. B., CHRIEN, K. S. and EVANS, L. B., *Chem. Eng.*, 1982, **89**, 125–7.
51. HALL, R. S., MATLEY, J. M. and MCNAUGHTON, K. J., *Chem. Eng.*, 1982, **89**, 80–116.
52. MULET, A., CORRIPIO, A. B. and EVANS, L. B., *Chem. Eng.*, 1981, **88**, 145–150.
53. WU, F. H., *Chem. Eng.*, 1984, **91**, 74–80.
54. BROWNELL, L. E. and YOUNG, E. H., *Process Equipment Design*, 1959, John Wiley & Sons, New York.
55. EVANS, F. L., *Equipment Design for Refineries and Chemical Plants*, 1974, Gulf Publishing Co., Houston.
56. CORRIPIO, A. B., CHRIEN, K. S. and EVANS, L. B., *Chem. Eng.*, 1982, **89**, 115–8.
57. RYANS, J. L. and CROLL, S., *Chem. Eng.*, 1981, **88**, 72–90.
58. LUDWIG, E. E., *Applied Process Design for Chemical and Petrochemical Plants*, Vol. 1, 1964, Gulf Publishing Co., Houston.
59. JACKSON, D. H. *Chem. Eng. Prog.*, 1948, **44**, 347–52.
60. RENSHAW, T. A., SAPAKIE, S. F. and HANSON, M. C., *Chem. Eng. Prog.*, 1982, **78**, 33–40.
61. ANON., *Thermal Handbook*, 1969, Alfa-Laval, Västeras.
62. KUMANA, J. D., *Chem. Eng.*, 1984, **91**, 169–72.

APPENDIX: DESIGN AND COST ESTIMATION OF THE MAIN EQUIPMENT OF MEFFE SYSTEMS

Each effect of the evaporation unit shown in Figs 3 and 4 consists of an FFE, a liquid–vapour separator and a centrifugal (recycle and feed) pump. Vacuum operation of the whole system is then assured by means of a barometric condenser and a series of ejectors and intercondensers.

This appendix will help to collect all the data necessary to estimate the investment and operating (see Section 5) costs of the equipment mentioned above.

FFE

Such equipment can be regarded as a shell-and-tube heat exchanger, where straight tubes as long as 10–12 m are secured at both ends in tubesheets welded to the shell. In order to prevent the high stresses resulting from differential thermal expansion between the tubes and shell, an expansion joint of the flanged-and-flued type in the shell may sometimes be provided.

Although some cost estimation procedures (see, for instance, ref. 49) take into account a great number of construction details of the heat exchanger (i.e. shell diameter, number and length of tubes, types of heads, shell and tube passes, etc.), which are actually determined in Section 4, it was found that the simple correlation of cost *vs* heat transfer surface can be sufficiently accurate for preliminary study-grade ($\pm 30\%$) cost estimates.[13,50,51] Therefore, the f.o.b. cost of fixed-head, stainless steel type 316 heat exchanger, rated for operating pressures smaller than 700 kPa and with total heat transfer surface (A) in the range 14–1100 m², can be evaluated as follows[52]

$$C_E = \frac{I_{CEFE}}{I^*_{CEFE}} \beta_m \beta_D C_E^* \tag{A.1}$$

with

$$\beta_m = 1 \cdot 414 + 0 \cdot 233 \, (\ln A) \tag{A.2}$$

$$\beta_D = \exp \{-0 \cdot 9 + 0 \cdot 091 \, (\ln A)\} \tag{A.3}$$

$$C_E^* = \exp \{8 \cdot 2 + 0 \cdot 051(\ln A) + 0 \cdot 068(\ln A)^2\} \tag{A.4}$$

where β_m is the cost correction factor for stainless steel type 316; β_D is the cost correction factor for fixed-head heat exchangers; I_{CEFE} is the *Chemical Engineering*'s Fabricated Equipment Index used to update the original cost data, which were referred to the first quarter of 1979 ($I^*_{CEFE} = 252 \cdot 5$); and C_E^* is the base cost for floating-head shell-and-tube exchangers in carbon steel and designed for a pressure of 700 kPa. The total installed cost of such equipment can finally be estimated by using Guthrie's[13] exchanger cost factor equal to $\zeta_E = 3 \cdot 29$.

When the design strategy of identical heat transfer surfaces is applied, the engineering costs are to be referred to a single FFE only, thus enabling the total installed cost of all the other $(n - 1)$ evaporators to be evaluated by using a properly reduced total module cost factor (i.e. $\zeta_E = 2 \cdot 53$).

Liquid–Vapour Separator

The design of such gravity settlers (consisting of a conical connector with a slope of 60°, a cylinder with diameter (D_S) and height (H_S) and an ellipsoidal head with minor axis equal to 1/4 of the separation diameter) can safely be based on a recommended vapour velocity equivalent to the terminal settling velocity (v_t) of liquid droplets with size (d_p) smaller than 250 µm[53]

$$v_t = \left\{ \frac{4}{3C_D} g d_p \left(\frac{\rho - \rho_V}{\rho_V} \right) \right\}^{1/2} \tag{A.5}$$

where C_D is the drag coefficient of falling particles, usually given as a function of the particle Reynolds number (Re_p):

$C_D = 24/Re_p$ $10^{-4} < Re_p < 2$ (Stokes's law)

$C_D = 18 \cdot 5/Re_p^{0.6}$ $2 < Re_p < 500$ (Intermediate region)

$C_D = 0 \cdot 44$ $500 < Re_p < 2 \times 10^5$ (Newton's law)

with

$$Re_p = \rho_V v_t d_p / \mu_V \tag{A.7}$$

Such design procedure yields a set of v_t values in good agreement with those recommended by Heid and Casten[3] for vapour phase disengagement droplet size (d_p) varying almost linearly from 450 µm at 130 kPa to 800 µm at 1·2 kPa. Since minimum loss in temperature requires the pressure drop to be kept to a minimum, vapour velocities should not exceed 15 m/s.[3]

Once the vapour velocity has been determined, the diameter D_S of the settler drum can be sized as follows:

$$D_S = \{4V/(\pi \rho_V v_t)\}^{1/2} \tag{A.8}$$

while the height H_S can be evaluated by assuming a residence time of a few seconds for the vapour entering the vessel on condition that

$$D_S \leqslant H_S \leqslant 4D_S \tag{A.9}$$

In this way, the shell weight of the drum under study is calculable via

$$W_S = \pi D_S (H_S + \tfrac{1}{2}D_S + 0 \cdot 345 D_S) t_S \rho_m \tag{A.10}$$

where $\rho_m = 8000$ kg/m³ is the specific gravity of material and t_S is the shell thickness. Since such equipment is under external pressure and is apt to collapse because of elastic instability, an appropriate wall thickness has to be chosen so as that the 'critical' pressure (P_C), at which a long, thin cylinder without circumferential stiffening rings or with such rings spaced at or beyond the critical length will buckle, is four[54]–five[52] times the difference between the external (atmospheric) pressure and the design (vacuum) pressure in the vessel. This can be carried out by using either the classic iterative graphical procedure outlined, for instance, by Brownell and Young[54] and Evans[55] or the numerical one developed by Mulet et al.[52] and based upon the following expression of P_C

$$P_C = \frac{2 \cdot 6(t_S/D_{So})^{2 \cdot 5} E_m}{(H_S/D_{So}) - 0 \cdot 45(t_S/D_{So})^{1/2}} \qquad (A.11)$$

where D_{So} is the external diameter of the separator and E_m is the modulus of elasticity of the material of construction used ($E_m = 2 \times 10^{11}$ Pa for stainless steel).

The total shop-fabricated, f.o.b. cost of such equipment (including the cost of nozzles, manholes, supports and shop-prime paint) can be estimated as follows[52]

$$C_S = \frac{I_{CEFE}}{I^*_{CEFE}} \beta_m C^*_S \qquad (A.12)$$

where β_m is equal to 2·1 for stainless steel type 316 and C^*_S is the base cost of carbon steel drums referred to the first quarter of 1979 and calculable via

$$C^*_S = \exp\{8 \cdot 6 - 0 \cdot 2165(\ln W_S) + 0 \cdot 0458(\ln W_S)^2\} \qquad (A.13)$$

for shell weights ranging from 2210 to 103 000 kg.

According to Guthrie,[13] the installed vessel cost can be derived by multiplying C_S by a total module cost factor (ζ_S) of 4.23.

Centrifugal Pump
Each effect of the evaporation unit is equipped with a stainless steel centrifugal pump to allow a given weight flow rate $\{(1 + E)S\}$ of the liquor to be moved from the bottom of the vapour–liquid separator to the top of the FFE belonging either to another effect or to the same effect for $E \neq 0$, that is to raise the liquor and overcome pipe and fittings friction. Therefore, the total dynamic head can be safely assumed as equal to

$$H_p = 2L(\mu/\mu_w)^{0 \cdot 25} \qquad (A.14)$$

The required brake horsepower (P_B) is then

$$P_B = g(1 + E)SH_p/\eta_p \qquad (A.15)$$

with

$$\eta_p = 0 \cdot 885 + 0 \cdot 00824(\ln Q_V) - 0 \cdot 012(\ln Q_V)^2 \qquad (A.16)$$

and

$$Q_V = (1 + E)S/\rho \qquad (A.17)$$

where η_P is the pump efficiency and Q_V is the volumetric flow rate ranging from $1{\cdot}2 \times 10^{-3}$ to $0{\cdot}32$ m^3/s.[56]

The electric motor driving the pump in question can be simply sized by overestimating the brake horsepower by 15% (P_M in kW), while its power consumption can be evaluated as follows

$$N_p = P_B/\eta_M \qquad (A.18)$$

with

$$\eta_M = 0{\cdot}5094 + 0{\cdot}056(\ln P_B) - 0{\cdot}00182(\ln P_B)^2 \qquad (A.19)$$

where η_M is the efficiency of three-phase electric motors expressed as a function of P_B in the range 746–380 000 W.[56]

Finally, the overall cost of each pump including the base plate, driving coupling and electric driver can be derived as[56]

$$C_P = \frac{I_{CEPC}}{I^*_{CEPC}} \beta_m \beta_T C_P^* + \frac{I_{CEE}}{I^*_{CEE}} C_M^* \qquad (A.20)$$

with

$$\beta_T = \exp\{0{\cdot}7147 - 0{\cdot}051(\ln \psi) + 0{\cdot}0102(\ln \psi)^2\} \qquad (A.21)$$

$$C_P^* = \exp\{7{\cdot}2234 + 0{\cdot}3451(\ln \psi) + 0{\cdot}0519(\ln \psi)^2\} \qquad (A.22)$$

$$C_M^* = \begin{cases} \exp\{5{\cdot}1288 + 0{\cdot}1234(\ln P_M) + 0{\cdot}15374(\ln P_M)^2\} \\ \qquad\qquad \text{for} \quad 0{\cdot}75 < P_M < 5{\cdot}6 \text{ kW} \\ \exp\{4{\cdot}101 + 0{\cdot}8472(\ln P_M) + 0{\cdot}024(\ln P_M)^2\} \\ \qquad\qquad \text{for} \quad 5{\cdot}6 < P_M < 186 \text{ kW} \end{cases} \qquad (A.23)$$

$$\psi = Q_V(gH_p)^{\frac{1}{2}} \qquad (A.24)$$

where β_m is the cost correction factor equal to 2 for stainless steel type 316; β_T is the design-type cost factor for one-stage, 1750 r/min, horizontally-split case pumps with Q_V and H_p varying in the ranges $0{\cdot}016$–$0{\cdot}32$ m^3/s and 15–150 m, respectively, and $P_B < 186$ kW; C_P^* and C_M^*, both referred to the first quarter of 1979, are the base costs, respectively, for one-stage, 3550 r/min, vertically-split case cast iron pumps and for 3600 r/min totally enclosed, fan-cooled motors; I_{CEPC} and I^*_{CEPC} ($=270$) are, respectively, the actual and reference *Chemical Engineering*'s Pumps and Compressors Indices; I_{CEE} and I^*_{CEE} ($= 175{\cdot}5$) are the actual and reference *Chemical Engineering*'s Electrical Equipment Indices; and ψ is a size parameter combining the maximum values of volumetric flow rate (Q_V) and dynamic head (H_p).

Finally, the installed cost of such equipment is obtained from C_P by means of Guthrie's[13] total module cost factor ($\zeta_P = 3\cdot38$).

Vacuum System

As shown in Fig. A.1, the vapour leaving the nth effect flows through a direct-contact barometric condenser, where the largest fraction of condensables is removed; then, the resulting stream of air saturated with water vapour is vented to the atmosphere by using a vacuum producer, such as steam-jet ejectors, rotary blowers, liquid-ring pumps, etc. For further details on vacuum device selection, see Ref. 57.

Since in this specific case the minimum absolute pressure (P_n) is usually greater than 2 kPa, a two-stage ejector system with a barometric intercondenser is generally recommended.[57,58]

In order to estimate the investment and operating costs of the vacuum system described, the amount of non-condensables of the process (G), the cooling water required at the barometric condenser (W_1) and intercondenser (W_2), and the motive steam flow rate (V_E) for

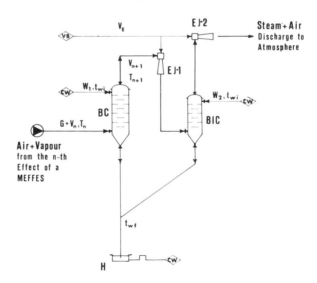

FIG. A.1. Typical scheme of the vacuum producer for the MEFFE systems shown in Figs 3 and 4. Equipment identification items: BC = barometric condenser; BIC = barometric intercondenser; EJ = steam-jet ejector; H = hydraulic seal. Utility identification items and stream symbols as in Fig. 3.

the two-stage ejector used are to be determined according to the following procedure.

(1) Estimate the amount of non-condensables of the process (G) by taking into account the air leakage occurring at piping and equipment connections and the air released from the inlet liquor. Among the alternative criteria used in practice (see Refs 3, 58), Jackson's method,[59] based on the summation of an average air leakage per each type of fitting, seems to be safe, thus yielding in this specific case about 4 kg air leakage per hour and for each effect (heat exchanger and liquid–vapour separator). Air coming in with the inlet stream (S_0) is usually negligible; however, it can be evaluated by assuming that the inlet liquor is saturated at the input temperature (t_0), thus involving about 2×10^{-5} kg air per kg liquor.[58]

The overall amount of air can therefore be evaluated as follows

$$G = 2 \times 10^{-5} S_0 + 1\cdot11 \times 10^{-3} n \qquad \text{(A.25)}$$

(2) Fix the temperature of the exit gaseous stream from the barometric condenser 3°C greater than the inlet cooling water temperature (t_{wi})

$$T_{n+1} = t_{wi} + 3 \qquad \text{(A.26)}$$

and set the outlet cooling water temperature (t_{wf}) as that of steam corresponding to vacuum less 3°C

$$t_{wf} = T_n - 3 \qquad \text{(A.27)}$$

(3) Calculate the cooling water consumption (W_1) by writing the heat balance across the barometric condenser

$$W_1 = \frac{(V_n - V_{n+1})(H_n - h_f^*) + (G c_G + V_{n+1} c_{wv})(T_n - T_{n+1})}{h_f^* - h_i^*} \qquad \text{(A.28)}$$

where

$$V_{n+1} = \frac{M_w}{M_a} \frac{p_s(T_{n+1})}{P_n - p_s(T_{n+1})} G \qquad \text{(A.29)}$$

and M_w and M_a are the molecular weight of water and air.

(4) Evaluate the motive steam requirement (V_E) as suggested by Ludwig[58]

$$V_E = z_1 z_2 z_3 (G + V_{n+1}) \qquad \text{(A.30)}$$

with

$$z_1 = 96 \cdot 766 - 18 \cdot 335(\ln P_n) + 0 \cdot 895(\ln P_n)^2 \qquad (A.31)$$

$$z_2 = 0 \cdot 995 + 0 \cdot 387(\ln x_G) + 0 \cdot 055(\ln x_G)^2 \qquad (A.32)$$

$$z_3 = \exp 0 \cdot 602 - 1 \cdot 092 P_m + 0 \cdot 387 P_m^2 \qquad (A.33)$$

where z_1 is the steam consumption factor (expressed as kg of motive steam per kg of air-vapour mixture), which depends on P_n (in Pa); z_2 is the non-condensable load factor, which is a function of the weight fraction (x_G) of the incondensables in the total mixture handled and z_3 is the steam pressure factor which is dependent upon the motive steam pressure (P_m—in MPa—at the nozzle sections of the ejectors examined). The above regressions (A.31–A.33) were derived from Ludwig's data[58] by means of the method of least squares and each of them exhibited a regression coefficient r^2 greater than 0·99.

(5) Estimate the cooling water required at the barometric intercondenser (W_2) in accordance with Ludwig's[58] suggestion to allow half of V_E to be entirely condensed

$$W_2 = \tfrac{1}{2} V_E \frac{H_E - h_f^*}{h_f^* - h_i^*} \geqslant 0 \cdot 6 \text{ kg/s} \qquad (A.34)$$

on condition that a minimum amount of cooling water is supplied to enable a safe and stable operation of the barometric intercondenser.[58]

(6) Estimate the total installed cost of the above vacuum system as indicated by Ryans and Croll[57]

$$C_V = \frac{I_{\text{CEPC}}}{I_{\text{CEPC}}^*} \beta_m C_V^* \qquad (A.35)$$

with

$$C_V^* = 1 \cdot 27 \times 10^5 V_E^{0 \cdot 35} \qquad (A.36)$$

where β_m is the cost correction factor equal to 1·1 for stainless steel type 316[57] and C_V^* is the base cost of a carbon-steel two-stage steam-jet ejector with barometric pre-condenser and intercondenser.[57]

Chapter 6

HEAT TRANSFER AND STERILISATION IN CONTINUOUS FLOW HEAT EXCHANGERS

CHRISTIAN TRÄGÅRDH and BERT-OVE PAULSSON
Division of Food Engineering, Lund University, Sweden

SUMMARY

Methods for the assessment of the characteristics of continuous heat exchangers for food sterilisation are lacking or underdeveloped in respect to both experimental and computational ways. This chapter introduces a new concept, thermal-time distribution (TTD), which increases the information of the operating characteristics of a sterilisation plant in respect to product quality changes. The development of high temperature, short time (HTST) and ultra high temperature (UHT) technologies has required, among other things, better methods for calculating the changes that the product has gone through while being heated. Theories describing the flow field and temperature field combined with models describing the kinetics of quality changes form the basis for the methods and give, as a result, the distribution of degree of quality changes that are a consequence of the different fluid particles being subjected to different temperature-time conditions. They are applicable to both laminar and turbulent flow. Demonstrations have been made for tube and scraped surface heat exchangers operating in turbulent flow, raising the product temperature from 100°C to 140°C. The demonstrations include kinetic models for quality changes.

NOMENCLATURE

a thermal diffusivity $(m^2 s^{-1})$
A area (m^2)
b, B coefficients

245

c concentration

c_p specific heat capacity ($J\,kg^{-1}\,K^{-1}$)

C coefficient

D_r decimal reduction time (min)

D_m mass diffusivity ($m^2\,s^{-1}$)

d diameter (m)

E distribution/frequency function

E_a activation energy ($J\,mol^{-1}$)

F distribution/probability function

J temperature jump ratio

k reaction rate coefficient (s^{-1})

k turbulent kinetic energy (eqn 18) ($J\,m^{-3}$)

$\dot{m}$ mass flux ($kg\,m^{-2}\,s^{-1}$)

N rotational speed (s^{-1})

n_r reaction order

n number of tanks

n_n viscosity exponent

$n(t)$ stochastic frequency function

$\dot{q}$ heat flux ($J\,m^{-2}\,s^{-1}$)

r reaction rate (s^{-1})

R gas constant ($J\,mol^{-1}\,K^{-1}$)

Re_a axial Reynolds number

t time (s)

T temperature (K)

Ta   Taylor number

$\mathbf{u}$ Eulerian velocity ($m\,s^{-1}$)

$\mathbf{v}$ Lagrangian velocity ($m\,s^{-1}$)

x Eulerian space coordinate (m)

$\mathbf{X}$ position of fluid particle (m)

α heat transfer coefficient ($J\,m^{-2}\,K^{-1}\,s^{-1}$)

γ shear strain (s^{-1})

ε turbulent energy dissipation rate ($J\,m^{-3}\,s^{-1}$)

η viscosity (Pa s)

κ Stoke–Einstein constant

λ thermal conductivity ($J\,m^{-1}\,K^{-1}\,s^{-1}$)

ρ density ($kg\,m^{-3}$)

τ shear stress (Pa)

Subscripts

i,j integer denoting spatial vector components

l integer denoting summation over all spatial directions

0   initial or reference value
r   rotor
w   exchange wall

Superscripts
'   turbulent fluctuating component
⁻   overbar, time averaged mean value
+   dimensionless

1. INTRODUCTION

Several approaches in continuous flow sterilisation of pumpable foods have been developed such as using a higher temperature and shorter time (HTST/UHT), processing highly viscous foods, processing liquid foods containing particulates and increased heat regeneration. The first of these developments (HTST/UHT) is a consequence of the increased knowledge of the kinetics of changes occurring in a product while being heated. The next two are a result of a desire to change from batch sterilisation to continuous processing which makes it possible to use the HTST/UHT technology. The last arose from the increased energy prices following the oil crisis in the early seventies. The technical solution of the larger heat regenerating section of sterilisation plants has often counteracted the trend towards the high temperature, short time concept.

A high temperature results in the reaction rates for bacteriological, chemical, biochemical and physical changes being fast. The importance of reliable kinetic data and methods to establish the thermal load that the product and its constituents are subjected to has increased correspondingly. Thus to be able to predict, or for a given process to establish, the thermal duty in respect to product quality a number of things must be known such as: (a) the kinetics of heat-induced product quality changes; (b) the thermal-time history of the individual fluid particles (result of fluid flow and heat transfer phenomena) and (c) the technical solution of the equipment and its operating characteristics.

This chapter will try to analyse existing equipment for continuous heat sterilisation in respect to their ability to fulfil the requirement of high product quality based on the HTST/UHT concept. The analysis will be based on available data using: (1) existing theories for kinetics of product quality changes; (2) governing conservation equations for momentum, heat and mass and (3) chemical reactor theories.

2. PRINCIPLES OF HIGH OR ULTRA HIGH
TEMPERATURE AND SHORT TIME

It is now well recognised that product quality changes have different rates and also that the temperature influence on the reaction rates differs. Commonly this is illustrated in a diagram of the type shown in Fig. 1. The significant feature here is the temperature dependence of the different reactions which results in the different inclination angles in Fig. 1. The lines in the figure represent operating lines where, in respect to each change, the same thermal treatment is achieved. It is defined as (if the temperature influence of the reaction rate follows the Arrhenius relationship and a first order reaction takes place)

$$ t = \frac{\ln \dfrac{c_0}{c} \exp (E_a/RT)}{k_0} \tag{1} $$

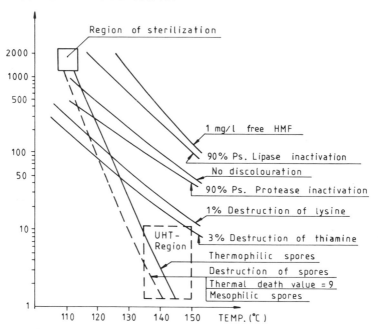

FIG. 1. Time–temperature relationships for different product changes taking place in milk during heating.

or

$$\ln t = \ln \left(\frac{\ln \dfrac{c_0}{c}}{k_0} \right) + E_a/RT$$

Thus each line represents a constant value of c_0/c for the actual constituent under whatever thermal-time history it is exposed to. The activation energy (or whatever it may be interpreted as, depending on what kind of reaction is considered) gives its inclination angle. By choosing an appropriate temperature–time relationship it is possible to favour certain product changes in relation to others. In the example in Fig. 1, which relates to changes occurring in the sterilisation of milk, following the line for constant spore destruction ratio, which is a necessity for the process, towards higher temperature and shorter time the result is less protease and lipase inactivation, lower lysine and thiamine destruction and lower content of hydroxymethylfurfural. Both desirable and undesirable product changes are favoured here when going from a low temperature and long time to higher temperature and shorter time. Most commonly for most products, it is claimed that the over-all effect as regards the total quality of the product is favoured by the HTST/UHT technique.[1] This is studied in more detail in the following section.

However, for physical reasons it is impossible to follow the HTST/UHT principle absolutely in a technical sterilisation plant. Heating and cooling courses often mean that the product is exposed to non-HTST/UHT conditions in certain instances and naturally the deviation is dependent on how rapid the temperature change is. The fluid particles are exposed to a broad spectrum of the thermal–time treatment as a result of the flow and temperature conditions in the equipment. These circumstances lead to a product experiencing a distribution of 'equivalent heating times' and often pure HTST/UHT conditions are not achieved. In this section on heat transfer and fluid flow phenomena a more detailed description is given on how things could be predicted and explained. In the last section, attempts to apply these theories to available equipment will be made.

3. KINETICS OF PRODUCT QUALITY CHANGES

Investigations on the kinetics of thermally induced product quality changes are widely found in food science literature. In their mathematical representation, and in the interpretation, the similarity to chemical and biochemical reaction kinetics is obvious. As in most cases, here we deal with complex reaction systems where a series of reactions both in series and in parallel take place and it is better to consider the kinetic modelling as data fitting procedures. Three models will be presented here, which differ in mathematical form and thus in agreement between themselves. The discrepancies between presented models to consider temperature and temperature effects on the same quality change seems to make them unsatisfactory and dubious. As mentioned, the demand for accurate predicting-methods grows as a consequence of the fast reaction rates caused by a higher process temperature.

3.1. Kinetic Models Theory

3.1.1. The Effect of Concentration
Following the definitions made in chemical reaction kinetics the processes are classified according to the number of atoms or molecules whose apparent concentration and interaction are determining the rate.

(a) Zero order reaction where the reaction rate of the concentration of the product quality is formulated as

$$\frac{dc}{dt} = -k \tag{2}$$

(b) For a n_r-th order reaction the reaction rate is

$$\frac{dc}{dt} = -kc^{n_r} \tag{3}$$

Common values for the reaction order n_r are 1 or 2. The reaction rate constant k is a function of the temperature and the concentration of chemical components, such as H^+, H_2O or Na^+, in the product. This model is often used when describing the inactivation of vitamins, enzymes and micro-organisms. The reaction order $n_r = 1$ is the most widely used for the inactivation of micro-organisms. In most cases it is convenient to use the integral form

$$\int_{c_0}^{c} \frac{dc}{c} = -\int_{0}^{t} k \, dt \tag{4}$$

or in the case of constant temperature

$$\ln \frac{c}{c_0} = -kt \tag{5}$$

Equation (5) can be re-written in the 10-logarithmic form

$$\log_{10} \frac{c_0}{c} = \frac{k}{2 \cdot 303} t$$

The D_r value (decimal-reduction-time) is the time necessary, at a specific temperature, to reduce the concentration to $1/10$ of the initial value. The D_r value is connected to the reaction rate coefficient in the following way

$$D_r = \frac{2 \cdot 303}{k} \tag{6}$$

3.1.2. The Effect of Temperature

The reaction rate dc/dt is always a function of the temperature. This means that the reaction rate constant k is a function of the temperature, as well.

The following equation can be derived from the gas law (reaction kinetics of Arrhenius and van't Hoff) taking the molecular activation energy E_a into account

$$k = k_0 \exp(-E_a/RT) \tag{7}$$

R is the gas constant and T is the temperature in °Kelvin.

Another way of modelling the temperature dependency of k is to use the Q value. This is defined as

$$Q_{10} = \frac{k_{T+10}}{k_T}$$

The Q_{10} value shows how much faster a reaction takes place when the temperature is raised by 10°C. As an alternative the Q value can be expressed by way of the Z value, the definition of which is

$$Z(°C) = \frac{10}{\log Q_{10}}$$

3.1.3. The Simultaneous Effects of Concentration and Temperature

Calculation of the reaction rate of a system with a time-dependent temperature is made possible using a combination of equations from

3.1.1. and 3.1.2. Reaction rates of micro-organisms can also be described using the model of Casolari.[2] For a system with an initial concentration c_0 of micro-organisms, the concentration of time t can be predicted by

$$\log c_t = (1 + k_c t)^{-1} \log c_0 \tag{8}$$

where k_c is

$$k_c = \exp\left(103 \cdot 7293 - \frac{2E_a}{RT}\right)$$

E_a, the activation energy value, is a function of the micro-organisms and their envirionment (H^+, Na^+, water activity, etc.).

3.2. Kinetic Model Applications

Two main problems are involved when using models in practice. Firstly the numerical value of the constants, and secondly, how well the models approximate reality.

The numerical value of a constant determined in one system is a function of its specific state regarding parameters such as chemical compound concentrations, water activity, etc. It is hazardous to apply values from one system to another where the conditions are not the same.

The second problem can be illustrated using the following example. A system of *Bacillus stearothermophilus* in water having a concentration of either 10^{10} or 10^5 has its temperature instantaneously increased from 30°C to 110°C. Thereafter the temperature is increased by 5°C/min until 135°C, the end point, is reached.

According to 3.1, there exist three different calculational methods: D_r and Q value (method 1), k and E_a (method 2) and the third method, that of Casolari. The numerical values of constants needed for the respective methods are

(1) $D_{121} = 1 \cdot 05$ (min) $Q_{10} = 2 \cdot 234$

(2) $k_0 = 9 \cdot 516 \times 10^{15}$ (min^{-1}) $E_a = 1 \cdot 18 \times 10^5$ (J min^{-1})

(3) $E_a = 173 \cdot 56 \times 10^3$ (J mol^{-1})   (based on minutes)

The constants were obtained from Malmborg.[3]

The results for each method having an initial concentration of 10^{10} are shown in Fig. 2a. Figure 2b gives the results for a concentration of 10^5. The calculated results differ significantly from each other. This is

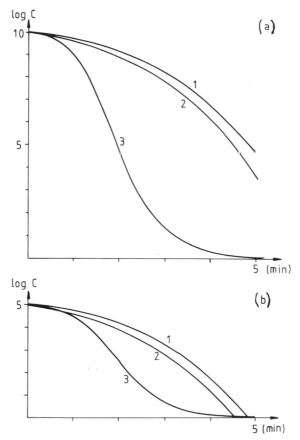

Fɪɢ. 2. Thermal inactivation of micro-organisms. Computed values using different inactivation models (1) the D–Z method, (2) the Arrhenius type of method and (3) the method of Casolari for two initial concentrations (a) 10^{10} micro-organisms per ml and (b) 10^5 micro-organisms per ml.

an effect of the models and not of the constants. It is open to question which model approximates reality best. The only possibility of choosing a convenient model is to use experimental data for comparison. Due to the problem above it is better to use a computer for the calculations as this permits faster testing of constants and models.

Table 1 gives some examples of the numerical values of some constants for different food properties.[6]

TABLE 1
KINETIC DATA FOR THE TEMPERATURE INDUCED CHANGES OF SOME
FOODS[6]

Chemical changes	$Z(°C)$	$E_a(KJ/mol)$	$D_{121}(min)$
Non-enzymatic browning	17–39	100–250	0·4–40
Fat oxidation		40–100	
Denaturation of proteins	5–10	250–800	5
Vitamin destruction			
In general	20–30	80–125	100–1000
Thiamine	20–30	90–125	38–380
Ascorbic acid (vitamin C)	51	65–160	245
Enzyme inactivation			
In general	7·55	40–125	1–10
Perioxidase	26–37	67–85	2–3
Micro-organisms			
B subtilis	6·8–13	230–400	0·4–0·76
C botulinum	8–12	265–340	0·1–0·1
C. thermosaccharolyticum	1·7–10	1340–1780	3–22

4. HEAT TRANSFER AND FLUID FLOW PHENOMENA

4.1. Heat Transfer

The presentation here will be restricted to those cases where the heat
transfer takes place through molecular diffusion (heat conduction),
eddy diffusion (turbulent exchange) and bulk diffusion (large scale
convective flow) as the commercially available equipment for continuous heat sterilisation uses these processes only. Microwave dielectric, electrical resistive heating, etc., are also technically possible
but there are no signs of these techniques commercially apart from
solid foods processing.

The length scale within each of these processes differs and so does
the heat transfer rate whereby energy is transported from one point to
another. Molecular diffusion, which normally has the lowest transfer
rate, works at molecular level as its physical nature is molecular
motion. Eddy diffusion, which here implies turbulent exchange by the
whole range of eddy sizes existing in a turbulent flow, has a small
length scale (the so-called Kolmogoroff's microscale) and a larger
length scale using the height of the flow channel. In spatial dimensions
for commonly appearing turbulent flows, this means a range from
10^{-4} m to 10^{-2} m. The heat transfer from the dimension $10^{-4}–10^{-10}$ m

is caused by molecular diffusion. In the case of streamline or laminar flow the heat penetration is controlled by the slow molecular diffusion only for the whole range of 10^{-2}–10^{-10} m.

Normally the eddy diffusion is several orders of magnitude larger than the molecular diffusion and it can be seen that turbulent flow conditions significantly enhance the possibilities of working according to the HTST/UHT principles through a more evenly distributed temperature across the flow channel combined with a possibility for rapid heating (or cooling).

The energy equation, or by another name the conservation equation of heat (enthalpy), describes these relations in mathematical terms, even though it does not directly relate to the different length scales. The time averaged energy equation for turbulent flow reads

$$\frac{D\bar{T}}{Dt} = -\frac{\partial}{\partial x_l}\left(\overline{T'u_l'} - a\frac{\partial\bar{T}}{\partial x_l}\right) \qquad (9)$$

$$\begin{array}{ccc} \text{bulk} & \text{eddy} & \text{molecular} \\ \text{diffusion} & \text{diffusion} & \text{diffusion} \end{array}$$

$$\frac{D\bar{T}}{Dt} = \frac{\partial\bar{T}}{\partial t} + \bar{u}_l\frac{\partial\bar{T}}{\partial x_l}$$

where D/Dt stands for $\partial/\partial t + u_l(\partial/\partial x_l)$. Correspondingly it is possible to formulate the heat flux resulting from the mentioned diffusion processes

$$\dot{q}_l = \rho c_p \bar{T}\bar{u}_l + \rho c_p \overline{T'u_l'} - \lambda\frac{\partial\bar{T}}{\partial x_l} \qquad (10)$$

$$\begin{array}{cccc} \text{total} & \text{bulk heat} & \text{eddy heat} & \text{molecular} \\ \text{heat} & \text{flux} & \text{flux} & \text{heat flux} \\ \text{flux} & & & \end{array}$$

In these equations it is understood that the subscript l means that the actual denominator should be summarised over the space coordinates so that for instance

$$\dot{q}_{tot} = \dot{q}_l = \dot{q}_1 + \dot{q}_2 + \dot{q}_3$$

The scalar equation T and vector quantity u is here, as is commonly done, divided into resp. time mean value and turbulent fluctuating part

$$u_i = \bar{u}_i + u_i'$$
$$T = \bar{T} + T' \qquad (11)$$

The time mean values of the fluctuating parts are by definition zero

$$\overline{u_i'} = 0$$

$$\overline{T'} = 0$$

but not the products

$$\overline{u_i' u_i'} \neq 0$$

$$\overline{u_i' T'} \neq 0$$

In the case of laminar flow the fluctuating components vanish and the instantaneous values correspond to the time mean values. The equations are reduced to

$$\frac{DT}{Dt} = \frac{\partial}{\partial x_l}\left(a\frac{\partial T}{\partial x_l}\right) \tag{12}$$

bulk   molecular
diffusion   diffusion

$$\frac{DT}{Dt} = \frac{\partial T}{\partial t} + u_l\frac{\partial T}{\partial x_l}$$

$$\dot{q}_l = \rho c_p T u_l - \left(\lambda\frac{\partial T}{\partial x_l}\right) \tag{13}$$

total   bulk   molecular
heat   heat   heat flux
flux   flux

and as expected the eddy diffusion disappears.

The heat flux $\dot{q}$, which is a vector, is strongly influenced by the flow conditions (fluid flow will be discussed later). In channel flow, which is the dominating system for heating of liquid foods, the heat flux thus differs depending on where in the flow channel it is considered and on the heat flux to the heat transfer surface.

Figure 3 shows the heat flux vector and its components for a typical heat transfer situation of turbulent and laminar heat transfer in a circular pipe evenly heated from the pipe wall. Turbulent flow conditions thus favour the possibility of operating according to HTST/UHT principles, as firstly they allow a higher heat transfer rate and secondly they reduce radial or cross-sectional temperature gradients due to the intensive turbulent exchange. The mean velocity profile is also in this respect more favourable in general for turbulent flow.

The thermal conductivity λ is dependent on, apart from the

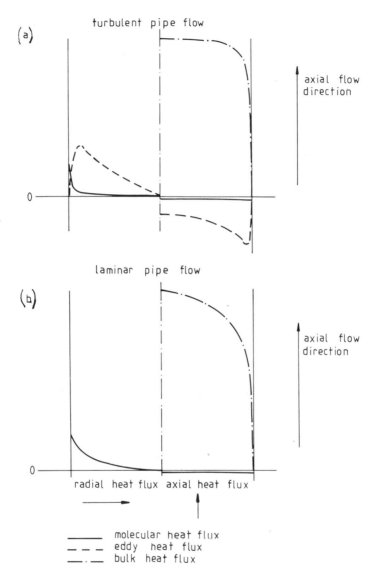

FIG. 3. Heat transfer rate in (1) radial direction and (2) axial direction for (a) turbulent and (b) laminar pipe flow with heat supply from the tube wall.

composition of the fluid, the temperature. Examples of data fitting models are found in the review by Cuevas and Cheryan.[4]

$$\lambda = b_0 + b_1 . T \tag{14}$$

or

$$\lambda = b_0 + b_1 . T + b_2 . T^2 \tag{15}$$

The temperature influence on the thermal conductivity is positive which means that during the heating course the heat supply is facilitated and the temperature gradient near the wall is lower. The method for solving these non-linear partial differential equations will not be dealt with here as they have been presented elsewhere (e.g. Ref. 19). They form the basis for calculating the temperature field on which the kinetics of product changes depend. As can be seen they contain the velocities of the flow domain which means that to be able to calculate the heat transfer phenomena the flow field must be known.

4.2. Fluid flow
As has been seen, the heat transfer conditions are almost totally governed by the fluid flow behaviour in the heat transfer equipment. The thermal properties of the liquid food that influence the heat flux are the thermal conductivity and the specific heat capacity and they vary within very restricted limits. In cases where the molecular diffusion has a magnitude close to or larger than the others the heat flux could of course be affected by the temperature gradient. Such conditions exist close to a solid wall or in laminar flow.

Conservation equations similar to the energy equation but for momentum here describe the flow field. Again they are presented in the time averaged form and here are called the Reynolds's equations

$$\rho \frac{D\bar{u}_i}{Dt} = -\frac{\partial \bar{p}}{\partial x_i} + \frac{\partial}{\partial x_l}\left(\eta \frac{\partial \bar{u}_i}{\partial x_l} - \rho . \overline{u_i' u_l'} \right) \tag{16}$$

molecular   molecular   momentum exchange
exchange   momentum   due tò the eddies
due to
convection

$$\frac{D\bar{u}_i}{Dt} = \frac{\partial \bar{u}_i}{\partial t} + \bar{u}_l \frac{\partial \bar{u}_i}{\partial x_l}$$

For laminar flow the conservation equations are, as a result of fluctuating components being non-existing and thereby zero

$$\rho \frac{Du_i}{Dt} = -\frac{\partial \rho}{\partial x_i} + \frac{\partial}{\partial x_l}\left(\eta \frac{\partial u_i}{\partial x_l}\right) \tag{17}$$

$$\frac{Du_i}{Dt} = \frac{\partial u_i}{\partial t} + u_l \frac{\partial u_i}{\partial x_l}$$

These non-linear partial differential equations are known as the Navier–Stokes equations.

Information on the Reynolds stresses $\overline{u_i'u_j'}$, the $\overline{u_i'T'}$ and $\overline{u_i'c_A'}$ is essential to be able to solve the conservation equations. Methods to solve this closure problem have been under development for approximately 100 years but exact solutions are still not available. Developed models vary in physical interpretation and description of turbulence phenomena and are always, at least in respect to those available for practical use, based on more or less strict approximations. The so-called k–ε model is one which during the last decade has been widely used and, thus, of which much experience has been gained. The k here stands for the turbulent kinetic energy and represents a measure of the turbulence intensity. The ε stands for the energy dissipation rate and represents a measure of the turbulence length scale. Together they then give the Reynolds stresses as

$$-\overline{u_i'u_j'} = B\rho k^2/\varepsilon \left(\frac{\partial \bar{u}_i}{\partial x_j}\right) \tag{18}$$

This belongs to a group called two-equation models as k and ε are achieved each from a conservation equation similar to the other conservation equations. The reader is referred to the article by Wolfgang Rodi[5] and the book by Hallström, Skjöldebrand and Trägårdh[6] for further details on turbulent modelling and its use in heat transfer and sterilisation calculations.

The fluid property that has a strong influence on the flow field is viscosity. First because in many cases it determines the flow condition. Strong viscous forces compared to the inertial forces stabilise the flow and when the stabilising effect is strong enough the flow becomes laminar. This is related to the magnitude of the dimensionless ratio of Reynolds

$$\mathrm{Re} = \frac{\rho u d}{\eta}$$

There is no distinct transition point between the laminar and turbulent flow regimes as the stability is influenced by a number of other physical factors like density and density gradient, physical obstacles, vibrations, etc. Secondly deviation from a constant viscosity, caused by a non-Newtonian behaviour and a temperature dependency, influences the velocity profile especially in the boundary layer. This is valid for both laminar and turbulent flow (when viscous forces influence the momentum exchange). As the heat flux expressions incorporate velocity components, the rheological properties in respect to heat sterilisation are of importance and as both the thermal diffusivity and molecular viscous diffusion are influenced by the temperature, the Navier–Stokes and energy equations must be solved (numerically) simultaneously (and iteratively). The temperature dependency of the viscosity is often modelled according to the Arrhenius type of expression

$$\eta = \eta_0 \exp\left(-B/T\right) \tag{19}$$

A non-Newtonian behaviour for food liquids is a result of the physical structure of the product. Constituents like macromolecules, polymers, particles, fibres, etc. interact with the shear flow and result in a viscosity which is most often dependent on the shear rate.

Examples of models for fluids without a yield stress are as follows.

The Power law or Ostwald–de Waele model

$$\tau_{ij} = \eta_0 \left(\frac{1}{2} \left| \sum_i \sum_j \gamma_{ij}\gamma_{ij} \right| \right)^{n_n-1} \gamma_{ij} \tag{20}$$

The Ellis model

$$\tau_{ij} = B_0 + B_1 \frac{1}{2} \sum_i \sum_j \tau_{ij}\tau_{ij}^{n_n-1} \gamma_{ij} \tag{21}$$

The Eyring–Powell model

$$\tau_{ij} = \eta_0\gamma_{ij} + \frac{1}{B_0} \sinh^{-1}\left(\frac{1}{B_1}\gamma_{ij}\right) \tag{22}$$

In the Newtonian case we have the usual stress–strain relationship

$$\tau_{ij} = \eta\left(\frac{\partial u_i}{\partial x_j} + \frac{\partial u_j}{\partial x_i}\right) \tag{23}$$

or

$$\tau_{ij} = \eta\gamma_{ij}$$

Further models for other types of fluids, like viscoelastic, are presented in the book by Schowalter[7] while books/articles by Tschubik and Maslow,[8] Holdsworth,[9] and Jowitt et al.[10] present rheological data of liquid foods.

4.3. Influence of Fluid and Heat Transfer Phenomena on Product Quality Changes

The mass balance of any product quality which could be quantified, such as content of constituents, number of micro-organisms, colour, sensory properties, etc., can be written in mathematical form as

$$\frac{D\bar{c}_A}{Dt} = \frac{1}{\rho}\frac{\partial}{\partial x_1}\left(D_m \frac{\partial \bar{c}_A}{\partial x_1} - \overline{c'_A u'_l}\right) + r \tag{24}$$

$$\frac{D\bar{c}_A}{Dt} = \frac{\partial \bar{c}_A}{\partial t} + \bar{u}_l \frac{\partial \bar{c}_A}{\partial x_l}$$

which is the time averaged conservation equation for compounds in incompressible turbulent flow analogous to the energy equation presented earlier. The corresponding conservation equation for laminar flow reads

$$\frac{Dc_A}{Dt} = \frac{1}{\rho}\frac{\partial}{\partial x_1}\left(D_m \frac{\partial c_A}{\partial x_1}\right) + r \tag{25}$$

$$\frac{Dc_A}{Dt} = \frac{\partial c_A}{\partial t} + u_l \cdot \frac{\partial c_A}{\partial x_1}$$

The source term r here represents the rate of change of a quality A. Thus to follow the equations presented in Section 3 it could be written as

$$r = \frac{dc_A}{dt}$$

$$\frac{dc_A}{dt} = -kc_A^{n_r}$$

$$k = k_0 \exp\left(-B/T\right)$$

The diffusivity D_m is a measure of the mobility of the constituent A or the structure it is bound to. For macromolecules and particles of spherical shape the diffusivity is fairly well described by the Stokes–

Einstein equation

$$D_m = \frac{\kappa T}{6\pi\eta} \qquad (26)$$

For many food constituents which are of large molecular size or bound to structures (particles) the mass diffusivity is very low. In most flow systems relevant to continuous heat sterilisation the diffusivity term in the conservation equation for such constituents or, as we here have preferred to use a more general name, qualities could be neglected. This means that in the preceding discussion of fluid particles travelling through the system the approximation, that they retain their identity and not exchange mass with the surrounding, could be valid.

In order to predict the information about product quality changes which occur in a continuous flow steriliser a distinction must be made between laminar and turbulent flow conditions. The information is interesting and desirable not only because it is a mean value of the quality but its distribution reflects the different fluid pathways through the flow channels. The result is that the different parts have different degrees of thermal treatment, or expressed in terms of HTST/UHT principles, different equivalent times.

For laminar flow this is in principle no problem as the equation of motion describes exactly what is happening and the fluid 'particles' follow precisely the path described by the velocity vector. Thus, it is possible to calculate the concentration profile through the heat exchanger and at the outlet obtain a concentration (or thermal–time) distribution function as defined by Paulsson and Trägårdh,[11] for example. In the turbulent case the situation is different as the equations describe a time-averaged process and cannot consider the chaotic movement that the fluid is exposed to during its passage through the equipment. The fluid particles here follow unpredictable pathways resulting in their thermal–time histories being unpredictable. The method used to solve the problem is influenced by the statistical description of turbulence phenomena or in advanced studies based on spectral analysis. The model presented by Paulsson and Trägårdh[11] incorporates methods developed in the chemical reactor theory.

4.3.1. Combined Chemical Reactor Models for Establishing Thermal–Time Distribution of Turbulent Fluid in Heat Exchangers

The flow system here is divided into a number of compartments each modelled as an ideal reactor such as plug flow, tank reactor or a combination of both. The temperature in each compartment (reactor) corresponds to that of the corresponding location in the heat exchanger. The flows connecting the reactors are determined by flow field, both the bulk flow and eddy exchange. A statistical random walk (Markov chain) procedure is then applied to determine the paths of a sufficiently large number of fluid particles through and between the reactors. Each reactor has its own residence time distribution function stochastically determining the residence time for each particle in each reactor. As the temperature is known for each reactor a summation of the thermal effect for each particle is for instance made as

$$\log c_0/c = \sum \sum \Delta t \, k_0 \exp\left(-B/T\right) \tag{27}$$

Results from simulations for a pipe heat exchanger have been presented according to the method by Paulsson and Trägårdh.[11]

4.3.2. Lagrangian Formulation of Fluid Particle Motion to Establish Fluid Flow in Heat Exchangers

The basic idea here is the same as previous, namely to tag a fluid particle on its way through the system and determine its temperature–time history from which its quality change can be computed. The general equation describing the Lagrangian position $\mathbf{X}$ of a fluid particle is

$$\mathbf{X}(\mathbf{X_0}, t) = \mathbf{X_0} + \int_0^t \mathbf{v}(\mathbf{X_0}, t_1) \, dt_1 \tag{28}$$

or

$$\mathbf{X}(\mathbf{X_0}, t) = \int_0^t \left[\bar{\mathbf{u}}(x(\mathbf{X_0}, t_1), t_1) + \mathbf{v}'(\mathbf{X_0}, t_1)\right] dt_1$$

As both the Eulerian velocity $\mathbf{u}$ and the Lagrangian velocity $\mathbf{v}$ are stochastic variables of the flow field, the position $\mathbf{X}$ is also random whose mean value is the concentration of the constituent it represents. The problem here is to model the fluctuating Lagrangian velocity vector $\mathbf{v}$. As seen earlier the mean flow velocity vector $\mathbf{u}$ could be computed by turbulence models. These turbulence models also give the magnitude of Eulerian point correlations $\overline{u_i' u_j'}$ as well as the kinetic energy k and dissipation rate ε of the turbulence which are used in modelling $\mathbf{v}'$. There are many suggestions to solve this 'random walk' equation.[12-14] The common feature of these is that they use a linear

Markov process type to model the random contribution

$$\frac{d\mathbf{v}'}{dt} + \beta\mathbf{u}' = f(t) \tag{29}$$

of the fluctuating velocity to be used to calculate the actual particle trajectory using the Lagrangian equation

$$\frac{d\mathbf{X}}{dt} = \mathbf{v}$$

dt is a suitable time step. The thermal-time effect for each fluid particle and the distribution of them is calculated as in the preceding section (eqn (27)).

So far we have not seen any modelling of this type for continuous heat exchangers in relation to sterilisation so for the moment only theories can be proposed. Work is at present going on in this direction so in the future there will be reliable methods based on these hypotheses to assess continuous heat sterilisers. However the idea by Pope[14] seems relevant in this respect as he has developed a model to determine a Lagrangian two-time probability density function for the fluctuating Lagrangian turbulent velocities. It is based on an exact solution of the equations of motion and then coupled to the transport equations for the Reynolds stresses. As the equations of motion and heat are using these types of equations for their closure problems the concept seems relevant here. However, it then also reduces computational efforts compared to other diffusion models as well as automatically puts in statistical functions describing the local flow conditions.

4.3.3. Thermal-Time Distribution for Laminar Flow in Heat Exchangers

In laminar flow, fluid particle trajectories are comparatively easily computed as here streamline flow exists resulting in the equations of motion (eqn (17)) giving their pathways directly. No averaging technique is applied and the closure problem as in the turbulence case does not exist. Equation (28) has now been reduced in the case of steady flow

$$\mathbf{X} = \mathbf{X_0} + \int_0^t \mathbf{u}(x(\mathbf{X_0}))\,dt \tag{31}$$

and if a sufficiently large number of fluid particles distributed over the inlet area A is used, in combination with eqn (27) to compute the individual thermal-time history, the whole distribution is achieved.

5. MODELLING OF THE STERILISATION EFFECT FOR COMMONLY COMMERCIALLY AVAILABLE HEAT EXCHANGERS

The most frequently used types of heat exchanger for continuous sterilisation are: plate, tube and scraped surface heat exchangers. As these types of heat exchangers have different configurations the flow patterns among them are also different. Such differences in flow patterns are expressed in the distribution of residence time and thermal time.

The distributions in the cases of tube and scraped surface heat exchangers will be calculated below.

5.1. Computer Models

5.1.1. Tube Heat Exchanger
During turbulent flow in a tube both radial and axial mixing occurs. Because of this, each fluid element has its own residence time as described by a residence time distribution (RTD). If the tube is heated, temperature profiles develop along the axis and the radius. Thus an element of fluid has a temperature depending on its location in the tube geometry. The distribution of thermal time (TTD) expresses mixing and the temperature profiles as a simultaneous result of effect. A model for a tube heat exchanger is developed by Paulsson and Trägårdh.[11]

5.1.2. Scraped Surface Heat Exchanger

The following approximations have been made in order to be able to make a model of the scraped heat exchanger.

(1) The degree of mixing in the radial direction is great enough to prevent gradients of temperature and velocity to occur (a Taylor vortex pattern is developed).
(2) Physical units such as c_p, v, ρ and λ are true constants, i.e. uniform in the whole of the volume.
(3) The heat flux through the wall is constant.

At the entrance of the heat exchanger there is a temperature jump as a result of axial dispersion. An equation describing this has been given by Maingonnat and Corrieu[16] in the following form

$$J = 0 \cdot 8 - \exp{(-2 \cdot 33 \times 10^{-2}(\text{Ta}/\text{Re}_a))} \tag{32}$$

where

$$J = \frac{T_{\text{inlet annulus}} - T_{\text{entrance}}}{T_{\text{exit}} - T_{\text{entrance}}}$$

and

$$\text{Ta} = 2\pi \left(\frac{d_w - d_r}{d_r}\right)^{0 \cdot 5} \frac{N d_r \rho (d_w - d_r)}{4\eta} \tag{33}$$

$$\text{Re}_a = \bar{u} \cdot (d_w - d_r) \cdot \rho/\eta \tag{34}$$

The scraped heat exchanger accounted for is using a model of perfect mixers in series. The total volume of the mixers equals the volume of the heat exchanger. To calculate the number of mixers needed an energy-balance over the first mixer is used. The first of the mixers, i.e. one situated at the entrance, must have a quantity of energy Q supplied, for the temperature jump ratio, J, to be maintained.

The amount (Q) of energy is thus

$$\dot{Q} = c_p(J \, \Delta T)\dot{m} \tag{35}$$

$\dot{m}$ being the mass flow and ΔT the total difference in temperature between the entrance (before the temperature jump) and the exit. α is calculated using eqn (36),[16] which reads as follows

$$\frac{\alpha d_w}{\lambda} = 2 \cdot 9 \left(\frac{\rho N d_w^2}{\eta}\right)^{0 \cdot 54} \left(\frac{\eta}{\lambda c_p}\right)^{0 \cdot 33} \tag{36}$$

Doing this, that part of the total heating surface which is demanded to supply the amount of energy Q is given by

$$\text{Surface 1} = \frac{\dot{Q}}{\alpha \, \Delta T_w} \tag{37}$$

In the above equation ΔT_w is the difference in temperature between the wall and the fluid. ΔT_w is uniform at all points of the heat exchanger wall, provided the heat flux is constant.

As the mixing conditions were assumed to be equal anywhere in the heat exchanger, each mixer thereby is of the same volume, i.e.

$$\text{Mixer volume} = \frac{\text{Surface 1}}{\text{Total surface}} \times \text{Total volume} \tag{38}$$

Equation (32) gives the temperature in mixer one. The temperature of the subsequent mixers are calculated using eqns (38) and (35).

The distribution of residence times in a mixer is described by[17]

$$E(t^+) = \exp(-t^+) \tag{39}$$

and when integrated over the time is transformed to the expression

$$F(t^+) = \int_0^t \exp(-z/t)\, d\left(\frac{z}{t}\right) = 1 - \exp(-t^+) \tag{40}$$

where $t^+ = t/\bar{t}$. Equation (40) can be stated as

$$t = \bar{t} \ln\left(\frac{1}{1 - F(t^+)}\right) \tag{41}$$

$F(t^+)$ has a value between 0 and 1.

Predicting the RTD and TTD, single fluid elements are followed through the mixers. As a fluid element passes through a mixer, its residence time is calculated by substituting the expression $F(t^+)$ in eqn (41) with a random number between 0 and 1, picked from a uniform distribution.

Thus the residence time and the thermal time, belonging to a specific fluid element, can be calculated using the following equations

$$RT = \sum_{i=1}^{n} t_i \tag{42}$$

$$TT = \sum_{i=1}^{n} f(t_i, T_i) \tag{43}$$

n being the number of mixers. A great number of fluid elements must be treated in this manner to give the TTD and the RTD. An exemplification $f(t, T)$ is

$$f(t, T) = \ln\frac{c_0}{c} = k_0 \exp(-E_a/RT) \tag{44}$$

5.1.3. Plate Heat Exchangers

No computer model available is able to simulate the plate heat exchanger in respect of the thermal effect obtained. The flow pattern is rather complex both within each flow channel and between them if connected in parallel or in flow distribution system. So far no TTD is thus possible to achieve. Residence time distribution could be obtained experimentally but as earlier declared the RTD only serves as a rough guide to the thermal characteristics in respect to product quality changes. However as experimental results of residence time are the only available information these will be used in a comparative study between the three heat exchangers under consideration.

5.2. Comparative Results between Scraped Surface, Tube and Plate Heat Exchangers

5.2.1. The Residence Time Distribution for Tube, Plate and Scraped Surface Heat Exchangers

The integrated RTD (eqn (40)) is shown in Fig. 4 for the cases of (a) tube, (b) plate and (c) scraped surface heat exchanger. The data for the plate heat exchanger are obtained from an experimental configuration of two parallel series, each having four channels, followed by two parallel series of two channels each. Nassauer[18] determined the RTD of a plate exchanger. For further informaton see Fig. 67 in Ref. 18. The RTD of the tube and the scraped surface heat exchanger are predicted using the models presented above. The primary data can be found in Section 5.2.2. The tube heat exchanger has the smallest distribution of RT and the scraped surface heat exchanger the largest one. The distribution is a result of mixing phenomena and velocity differences in the equipment and, if taking place in parts of the heat exchanger where the temperature is high, the result is a corresponding

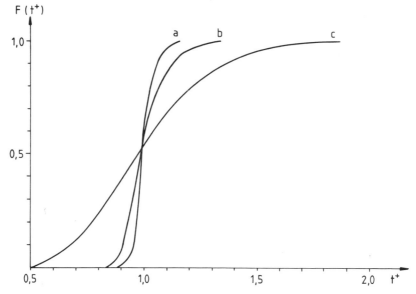

FIG. 4. Typical residence time distributions for a (a) tube, (b) plate and (c) scraped surface heat exchanger.

distribution of the thermal time effect. The RTD, under certain defined conditions, is a measure of the non-uniformity of the product treatment. Thus the smallest distribution of the residence times is generally desired, giving an even treatment of the product.

5.2.2. The Thermal Time Distribution for Tube and Scraped Surface Heat Exchangers

The computer simulations have been done using the following:

(a) The entrance temperature is 100°C and the exit temperature is raised between 110°C and 140°C.
(b) Mean residence time is 38 s.
(c) Values of physical constants used (c_p, ρ, λ, v, etc.) are those of water.
(d) The flux of heat is constant over the whole of the heating surface.
(e) The thermal time is predicted according to the Arrhenius equation

$$\text{TT} = \log \frac{c_0}{c} = \frac{k_0}{2 \cdot 303} \int_0^t \exp\left(-E_a/RT\right) dt \qquad (45)$$

These assumptions are used both for the tube and the scraped surface heat exchanger. The primary data for the tube heat exchanger are length = 7·0 m; diameter = 0·013 m and flow of mass = 0·023 kg/s. The primary data for the scraped surface heat exchanger are diameter (exchange wall) = 0·13 m; diameter (rotor) = 0·106 m; rotational speed = 4/s; flow of mass = 0·1 kg/s and length = 0·9 m. Different exit temperatures are achieved by variation of the flux of heat through the wall. The average value for the TTD at different exit temperatures has been predicted for the two cases of A and B. The constants belonging to A have been set to the following values: $k_0 = 7 \cdot 4 \times 10^7$ (min^{-1}) and $E_a = 52\,000$ (J/mol). For the case of B the constants have the values: $k_0 = 1 \cdot 4 \times 10^{13}$ (min^{-1}) and $E_a = 89\,000$ (J/mol).

Figure 5 shows the variation of the mean average of the TTD with the exit temperature (the dashed lines) for the cases of A and B. Figure 5 also shows for A and B the TTD ($E(\log c_0/c)$) for the tube (thick solid lines) and the scraped surface heat exchanger (thin solid lines) for two temperatures (120°C, 140°C). At all exit temperatures of the scraped surface heat exchanger the distribution of thermal time is larger when compared to the tube. In the case of B, having an exit

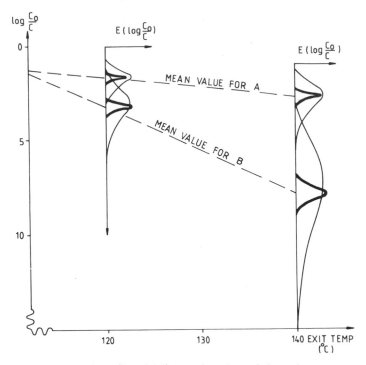

Fig. 5. The thermal time ($\log C_0/C$) as a function of the exit temperature for the two cases of $A(k_0 = 7\cdot4 \times 10^7, E_a = 52\,000)$ and $B(k_0 = 1\cdot4 \times 10^{13}, E_a = 89\,000)$. Average values of the TTD are represented by a dashed line. The TTD of the tube is represented at two different temperatures (120°C, 140°C) by a thick solid line, and the TTD of the scraped surface heat exchanger, also at the same two temperatures, by a thin solid line.

temperature of 140°C, the TTD continues from $\log c_0/c = 2$ to $\log c_0/c = 15$, this to be compared to the tube heat exchanger variation of between $\log c_0/c = 6$ and $\log c_0/c = 10$.

It is obvious that the scraped surface heat exchanger does not operate under HTST/UHT principles as calculated here. The broad thermal time distribution means that a part of the product could be treated out of this region as defined in Fig. 1. The distribution curves in Fig. 5 could be transformed into Fig. 1 by eqn (5). The tube heat exchanger however fulfils the requirement under the premises made here. The plate heat exchanger would probably fall in between these two.

6. CONCLUSIONS

The major aim of this chapter is to introduce the concept of thermal-time distribution instead of the residence-time distribution as one characteristic of a heat exchanger. The usefulness of such data is demonstrated using a conventional time–temperature diagram (Figs 1 and 5) and it can be seen how the design of the process and equipment fulfils the demand for an even product treatment within a desired operating region.

Experimental methods to obtain data are not at the moment available. As soon as the theoretical and computational tools described here become more developed even experimental procedures could be determined. The data presented here in the form of thermal–time distribution have one big flaw, they cannot be verified experimentally.

Also we conclude, according to Fig. 2, that there remains some controversy over how to model the kinetics of certain product quality changes. A proper model is one necessity towards a correct prediction of the achieved product quality.

ACKNOWLEDGEMENT

We want to thank Professor Bengt Hallström for his kind assistance during the preparation of this chapter.

REFERENCES

1. KESSLER, H. G., *Food Engineering and Dairy Technology*, Verlag A. Kessler, Freising, 1981.
2. CASOLARI, A., *J. Theoretical Biol.*, 1981, **88**, 1–34.
3. MALMBORG, A., In *Proceedings of the Sixth World Congress of Food Science and Technology*, ed. by J. V. McLoughlin and B. M. McKenna, Boole Press, Dublin, 1983.
4. CUEVAS, R. and CHERYAN, M., *J. Food Process Eng.*, 1978, **2**, 223.
5. RODI, W., *Turbulence models and their applications in hydraulics*, International Association for Hydraulic Research, Delft, 1980.
6. HALLSTRÖM, B., SKJÖLDEBRAND, C. and TRÄGÅRDH, CH., *Heat Transfer and Foods*, Elsevier Applied Science, 1985.
7. SCHOWALTER, W. R., *Mechanics of Non-Newtonian Fluids*, Pergamon Press, Oxford, 1978.
8. TSCHUBIK, I. and MASLOW, A. M., *Wärmephysikalische Konstanten von Lebensmitteln und Halbfabrikaten*, VEB Fachbuchverlag, Leipzig, 1973.

272   CHRISTIAN TRÄGÅRDH AND BERT-OVE PAULSSON

9. HOLDSWORTH, S. D., *Biochemistry,* 1969, **4,** 15–33.
10. JOWITT, R., ESCHER, F., HALLSTRÖM, B., MEFFERT, H. F. H., SPIESS, W. E. L. and VOS, G., (eds), *Physical Properties of Foods,* Applied Science Publishers, London, 1983.
11. PAULSSON, B.-O. and TRÄGÅRDH, CH., In *Proceedings from the 3rd International Congress on Engineering and Food, Dublin, 1983,* Applied Science Publishers, London, 1984.
12. CORRSIN, S., In *The Mechanics of Turbulence,* Gordon and Breach Science Publishers, New York, 1964.
13. LEE, N. and DUKLER, A. E., *A.I.Ch.E. J.,* 1976, **22,** 449–55.
14. POPE, S. B., *Phys. Fluids,* 1983, **26,** 3448.
15. HINZE, J. O., *Turbulence,* McGraw-Hill, New York, 1975.
16. MAINGONNAT, J. F. and CORRIEU, G., In *Proceedings of the 3rd International Congress on Engineering and Food, Dublin, 1983,* Applied Science Publishers, London, 1984.
17. RAO, M. A. and LONCIN, M., *Lebensm.-Wiss. u. Technol.,* 1974, **7,** 5–12.
18. NASSAUER, J., Thesis, Technischen Universität München, 1978.
19. PATANKAR, S. V., *Numerical Heat Transfer and Fluid Flow,* Hemisphere Publishing Co., New York, 1980.

Chapter 7

POTENTIAL APPLICATIONS OF FLUIDISATION TO FOOD PRESERVATION

Gilbert M. Rios

Laboratoire de Génie Chimique, Université des Sciences et Techniques du Languedoc, Montpellier, France

Henri Gibert and Jean L. Baxerres

GEPICA, Institut du Génie Chimique, Toulouse, France

SUMMARY

This paper makes a brief survey of the current state-of-affairs in fluidisation. It does not intend to cover all aspects of the subject but sets out to review the more attractive characteristics of fluidised contacting techniques, as well as to indicate their implications and limitations in the field of food processing. Newly patented devices are presented that overcome some difficulties and potential applications of fluidisation to food preservation are then discussed.

This paper has not been written with any particular audience in mind and different groups of readers may find it of interest. However because in several respects fluidisation for food processing is rather a technique of the 'future' than of the 'present time', the overriding consideration governing the choice of material is its relevancy to process analysis and design more than to immediate use for plant calculation. For this reason the introduction of new concepts and original ideas flowing from either the personal experience of authors or the results of other outstanding research works has been highlighted.

1. INTRODUCTION

Fluidisation is the operation by which solids are transformed into a fluid-like state through contact with a gas or a liquid. This method has a number of unusual characteristics and fluidisation engineering is concerned with taking advantage of these.

Since 1922, the date when the first patent for a fluidised bed was awarded, the technique has been extensively used in the chemical industry. However only a few applications of it are found in the field of food treatment. This seems chiefly due to the fact that up to recent years Food Engineering was an 'art' more than a 'science', much more concerned with product quality than with process improvement. At the present time the need for processes able to treat larger and larger quantities of products, as well as to save more and more energy, means that Chemical Engineering is a leading subject for Food Engineers. In particular, powerful aspects of fluidisation have begun to be perceived and people are taking more and more interest in the potential applications of the technique for food treatment.

The application of fluidisation principles in the food processing industry can be dated back to the early 1960s when fluidised-bed freezing was first developed. Then the concept of fluidised-bed cooking was conceived by Jowit and Whitney at the University of Reading in England in 1973. More recently other investigation programmes on 'Potential applications of fluidisation to food preservation' have been started elsewhere in the World. As far as we know, all the industrial developments of fluidisation in the food processing area and most of the outstanding research works are related to operations where solid particles are fluidised by a gas. However, from now on, one can also imagine new efficient processes where solids would be suspended in a continuous liquid medium and where either the solid or the liquid phase properties could be modified, which makes the field of applications for fluidisation techniques considerably wider. For convenience in what follows the first applications will be referred to as 'gas–solid' applications (even if operations are carried out in the presence of liquid droplets) and the others as 'liquid–solid' applications (even if gas bubbles are dispersed in the system).

This short paper is the first general review of the entire subject matter. It is divided into three major sections: the first section, in which general aspects and new developments of fluidisation techniques are introduced; and two other parts where some typical gas–solid and liquid–solid applications of fluidisation to food preservation are successively presented.

2. BASIC ASPECTS AND NEW DEVELOPMENTS IN FLUIDISATION

2.1. Basic Knowledge[1-5]

2.1.1. Fluidisation Phenomena

If a liquid or gas is evenly distributed at the bottom of a bed of solids and passes upward through it at a low flow rate, it merely percolates through the void spaces between stationary particles. This is a 'fixed bed' (Fig. 1).

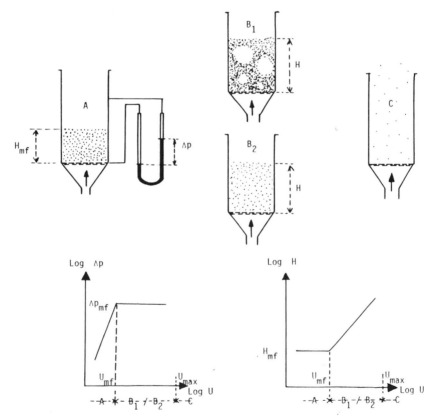

FIG. 1. Fluidisation phenomenon. A = fixed bed, B_1 = gas-fluidised bed, B_2 = liquid-fluidised bed, C = solid transport.

As the flow rate increases a point is reached where all the particles are just suspended in the upward flowing liquid or gas. At this point the frictional force between particles and fluid counterbalances the apparent weight of solids and the pressure drop through the bed equals the weight of fluid and particles per unit cross-sectional area. The bed is said at 'minimum fluidisation'.

In liquid–solid systems an increase in flow rate above this point usually results in a smooth homogeneous expansion of the bed. In gas–solid systems, beyond a short transition zone with an extent mainly depending on product properties, large instabilities with bubbling, channeling or slugging of gas are observed. In this case the bed looks very much like a boiling liquid with a continuous dense phase composed of gas and closely packed solid particles and a discontinuous bubble phase formed of large voids.

At a sufficiently high fluid flow rate the terminal velocity of particles is exceeded and solids are transported out of the column. The 'maximum fluidisation' working conditions have been reached.

As a general rule fluidisation can be operated over a wide range

$$10 < \frac{U_{max}}{U_{mf}} < 80 \tag{1}$$

where U (the 'superficial upflow velocity') is defined as the ratio of fluid flow rate to the column cross-sectional area. The lower limit corresponds to turbulent flow conditions and is approached when working with large particles and/or a liquid, while the upper value is more important in the gas fluidisation of fine particles with laminar flow conditions.

Reliable equations for minimum and maximum fluidisation velocity calculation, as well as constant bed pressure drop forecasting, may be found in the literature. As an example for spherical particles one can deduce U_{mf} values from the Ergun equation

$$Ga = 150 \frac{1 - \varepsilon_{mf}}{\varepsilon_{mf}^3} Re_{mf} + 1.75 \frac{1}{\varepsilon_{mf}^3} Re_{mf}^2 \tag{2}$$

where

$$Ga = \frac{\rho_F(\rho_S - \rho_F)g d_S^3}{\mu_F^2} = \text{Galileo number} \tag{3}$$

and

$$Re_{mf} = \frac{d_S U_{mf} \rho_F}{\mu_F} = \text{Reynolds number} \tag{4}$$

(In these equations d_S, ρ_S, ρ_F and μ_F refer to particle diameter, solid or fluid density and fluid viscosity, respectively.)

Taking ε_{mf} as 0·4, the equation simplifies to

$$Re_{mf} = 25 \cdot 7(\sqrt{1 + 5 \cdot 53 \times 10^{-5} Ga} - 1) \tag{5}$$

This equation can also be used for non-spherical particles if the diameter of a sphere with the same specific surface area as the particles is used.

The maximum fluidisation velocity that roughly corresponds to the free-falling velocity of particles may be estimated through

$$\begin{cases} Re_{max} < 0 \cdot 2 & Ga = 18\, Re_{max} & \text{(Stokes)} & (6) \\ 0 \cdot 2 < Re_{max} < 550 & Ga = 18\, Re_{max} + 2 \cdot 7\, Re_{max}^{1 \cdot 7} & \text{(Schiller and} \\ & & \text{Naumann)} & (7) \\ Re_{max} > 550 & Ga = 0 \cdot 33\, Re_{max}^2 & \text{(Newton)} & (8) \end{cases}$$

The pressure drop over the fluidised bed is given by

$$\begin{aligned} \Delta P &= (\rho_S - \rho_F)(1 - \varepsilon)Hg \\ &= (\rho_S - \rho_F)(1 - \varepsilon_{mf})H_{mf}g \end{aligned} \tag{9}$$

with H being the bed height.

Fluidisation of solid particles in the presence of more than one fluid has been also studied. The fluidised beds used in these studies were ordinarily established by co-current upward flows of a gas and a liquid as shown in Fig. 2, the liquid forming a continuous phase and the gas a discontinuous bubbling phase. Other schemes have been imagined with countercurrent flows of gas and liquid, using two immiscible liquids or even with a stationary liquid phase in which the particles would be suspended entirely as a result of bubble movement. Up to now they have been the subject of relatively little research due to the low number of direct applications in the chemical industry, but advantageous behaviour and characteristics have been discovered.

2.1.2. Advantages and Disadvantages

In the case of fixed beds there is no solid mixing. However, in the case of fluidised beds where the particles are mobile, mixing is important, particularly in the axial direction. The resulting fluid-like behaviour of solids with its rapid, easy and economic transport and handling and its intimate fluid contacting is the most important property of fluidised beds.

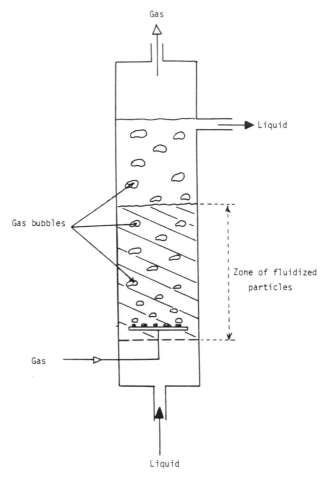

FIG. 2. Co-current three phase fluidised bed.

For batch treatments this mixing is helpful since it gives a uniform solid product. But it leads to scattered residence time distributions for continuous operations, especially when polydisperse materials with a range of different sized particles have to be treated. In that case plant designers must have recourse to baffling, multistage arrangements, conveyor belt systems (Fig. 3), etc., to warrant uniform product treatment conditions. Sometimes the stability of hydrodynamic conditions must be carefully looked at. In opposition to a widely accepted opinion, it must be also borne in mind that fluidised particles are ordinarily treated in a gentle way without damaging.

In a fluidised bed, there are also favourable conditions for rapid heat and mass transfer between the solids and the fluid because of rapid solid mixing and large heat exchange surfaces. Moreover the coefficients for the transfer of heat (and mass to a lesser extent) to boundary surfaces are very high when compared with other modes of contacting. Hence fluidised beds are often used both as heat exchangers and as chemical reactors, particularly where close control of temperature is required and where large amounts of heat must be added to or removed from the system.

These desirable or undesirable characteristics are not found to the same extent in gas–solid and liquid–solid systems: as a general rule better solid mixing characteristics, as well as higher wall-to-bed heat transfer coefficients and more uniform bed temperatures, are reached in gas fluidised beds. Moreover it must be kept in mind that many other factors may influence bed properties such as the column dimensions, the fluid flow rate, the type of distributor, etc.

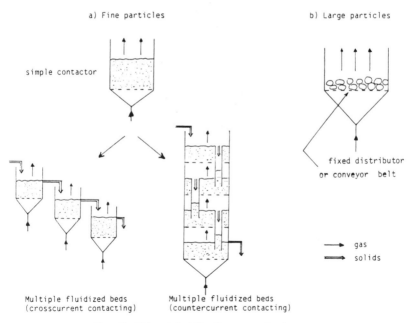

FIG. 3. Direct fluidisation: usual schemes.

2.1.3. Practical Limitations of Traditional Fluidisation

The ease with which particles fluidise and the range of operating conditions for fluidisation vary greatly among gas–solid systems. One major factor is the size of solids. In general fine particles less than 1 or 2 mm in diameter are easily fluidised, either in deep large beds or in shallow narrow systems, at either a low or a high flow rate. Larger solids fluidise poorly except when shallow beds are used with low multiples of the minimum fluidisation velocity. But in that case the energy consumption per unit mass of solid is high and the product

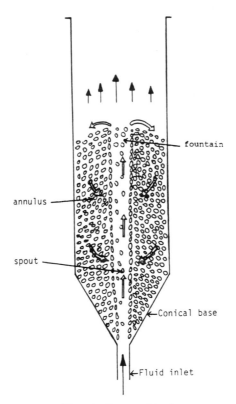

FIG. 4. Spouted bed.

quality is not necessarily good (due to bad solid mixing conditions and less attractive transfer properties); so the use of traditional fluidisation is questionable.

Sometimes people have recourse to a modified contacting scheme: the so-called 'spouted bed'. In that arrangement (Fig. 4) the fluid is injected vertically through a centrally located smooth opening at the base of the vessel, with a large enough rate to cause a stream of particles to rise rapidly in a hollowed central core within the bed of solids; then the particles rain back near the column wall, and they travel slowly downwards as a loosely packed bed through which gas percolates upwards. Unfortunately with such a device less uniform temperatures and treatment conditions are obtained.

When particles are immersed in a continuous liquid phase, the quality of fluidisation does not depend too much on the solid size, and good operating conditions may be reached even with coarse particles. But other limitations may arise from the fact that particles have nearly the same density as the liquid, or present a sticky surface, or from the fact that only a slight liquid retention is permitted which enhances the probability for solid caking at the top of the bed and proves to be detrimental to good phase contacting and related attractive properties.

2.2. New Contacting Systems Well Adapted to Food Treatment

2.2.1. Gas–Solid Systems

At first sight gas fluidisation appears to be a powerful technique for food processing owing to both the usual divided form of solid foodstuffs and the large number of unit operations bringing into contact solid particles and gas. Nevertheless in most cases coarse particles are to be treated and systems more efficient than ordinary beds need to be used. Recently two new devices have been proposed to improve phase contact efficiency in such cases: the 'whirling bed' and the 'fluidised floating cell'.

The 'whirling bed' as proposed by Baxerres and Gibert[6,7] is essentially a solid circulation system in which a high intensity mixing movement is obtained using a special distributor. Basically the idea is to prevent uniform gas distribution at the bottom of the column by fixing a wedge on the grid plate (Fig. 5). A cyclic movement of the solids results for superficial gas velocities higher than a minimum value referred to as 'minimum whirling velocity' and denoted U_{mw}. Unlike U_{mf}, this value does not vary too much as a function of solid

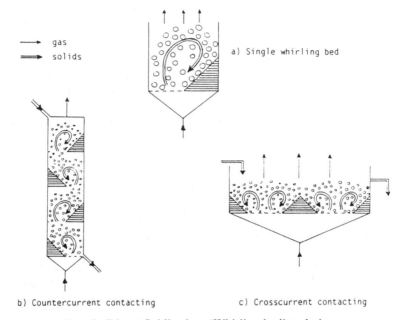

FIG. 5. Direct fluidisation: 'Whirling bed' technique.

properties; for particle sizes between 3 and 10 mm and solid densities between 1000 and 3000 kg/m^3, U_{mw} has been found roughly constant and nearly equal to 3·5 m/s.[7] Apart from gas velocity other requirements for smooth working conditions concern the bed height (as a rule the solid height at rest must not exceed the upper wedge level), the particle diameter (necessarily higher than 1 mm) and the particle to column size ratio (that must range between 10^{-1} and 10^{-2}). In the whirling state the bed pressure drop remains close to the weight of solid per unit cross-sectional area as in traditional systems and fluid-to-solid heat transfer coefficients as high as 300 W/m^2 °C are obtained that warrant uniform temperature conditions within the cell. It is also worth noting that easy-to-build multicell arrangements may be used either when large volume units are needed or when fairly uniform residence time distributions for continuous treatments are desired.

In a 'fluidised flotation cell' (Fig. 6a) light large particles—or 'objects'—are not directly suspended in a gas stream: they are buoyed by an intermediate fluidised medium composed of heavy fine particles or 'fines'.[8] The flotation of objects is governed by the vertical bed pressure distribution and mixing is induced by bubbling. As a general

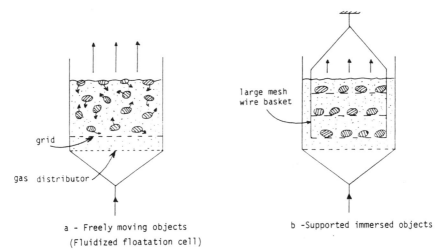

a - Freely moving objects   b -Supported immersed objects
(Fluidized floatation cell)

FIG. 6. Indirect fluidisation (two solid phases).

rule good working conditions with smooth and uniform treatment of products may be observed for object-to-bed density ratios between 0·5 and 0·9, object-to-bed volume ratios less than 0·4 and gas velocities two or three times above the minimum fluidisation fine velocity.[9] Otherwise objects tend to accumulate in fixed layers at the top of the bed (very light objects and/or low gas velocities) or to sink irreversibly (over-heavy objects and/or excessive loads) which is detrimental to treatment. To overcome these difficulties, one can change fines, vary gas flow rate or provide the apparatus with a well adapted conveyor system for transporting objects through fines (Fig. 6b). As indicated in Table 1, bed-to-object heat transfer coefficients between 200 and

TABLE 1

FLOTATION OF 8 mm POLYACETAL SPHERES ON FLUIDISED BEDS OF GLASS BEADS

Fine diameter (μm)	Minimum fluidisation velocity (cm/s)	Working velocity (cm/s)	Heat transfer coefficient (W/m²°C)
200–250	3·0	11·0	590
250–315	4·6	13·5	560
315–400	12·6	27·0	340
610–700	32·6	60·0	260
1200–1400	56·0	95·0	200

(Fine density: 2600 kg/m³ → fixed bed density: 1600 kg/m³. Object density: 1360 kg/m³. Object minimum fluidisation velocity: 1·5 m/s.)

600 W/m² °C   may   be   obtained   depending   on   fine   particle characteristics.[9,10]

2.2.2. Liquid–Solid Systems

The basic principles of whirling bed apparatuses, that is to say the simultaneous use of a horizontal fluid distributor plate responsible for the local upward entrainment of particles and of an inclined deflector well-adapted to bring the solid back to the fluid injection zone, have been extended to those cases where at least one phase is a continuous liquid phase. Experiments have been conducted in monocell and multicell plants (Fig. 7), with stationary or circulating liquids, by

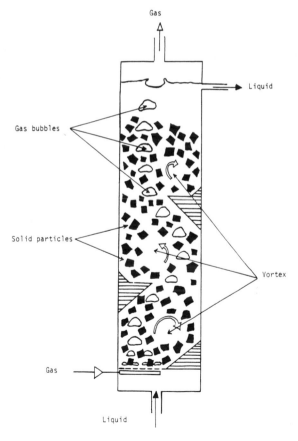

FIG. 7. Co-current three phase multicell whirling arrangement.

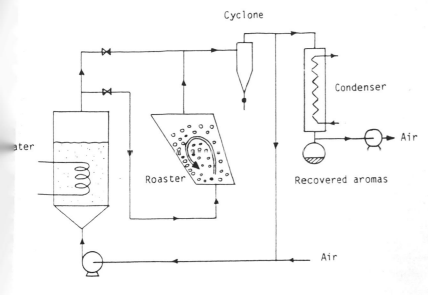

FIG. 10. Coffee roasting pilot plant.

unloading steps of the solids, the roasting chamber was disconnected from the entire circuit by quickly adjusting two on–off valves, so that hot gas could be kept within the apparatus, and its properties could approach those expected for a continuous operation in the long run.

Experiments clearly indicate that the most favourable working conditions are reached for bed temperatures between 200°C and 260°C; below the lower limit roasting times greatly increase; above the upper temperature value the physical aspect of roasted coffee grains as well as the organoleptic properties of beverages are damaged. The fluidised bed roasting times are considerably lower than those ordinarily observed with traditional apparatuses; they lie in the range between 2 and 4 min whereas they are found between 10 and 15 min for drum roasters! Roasted grains have attractive characteristics: weight losses are minimised (between 12 and 17% against 18–20% for other processes), swelling is increased and the whole physical aspect of grains is better; moreover the organoleptic characteristics of beverages are equal if not superior to the ones obtained with other roasting methods. Valuable aromatic compounds may be extracted from the exit stream with gas recirculation rates as high as 80%; apparently the thermal decomposition of 'fragile' organic compounds entrapped in

injecting a gas to promote movement. Various particles with a density higher or lower than the liquid one have been tested.[11,12]

It has been found that, when the liquid-to-solid density difference does not exceed 10%, the technique yields a perfect liquid mixing as well as a fairly uniform solid motion, even in those cases previously mentioned where hard fluidised contacting conditions prevail. Moreover it has been observed that the device guarantees high mass transfer rates between gas and liquid and facilitates liquid–solid reactions. Finally, from low gas flow rates needed to ensure good working conditions, it has been deduced that energy consumption is reduced.

3. GAS–SOLID APPLICATIONS

For treating food pieces larger than several centimetres in size, only immersion techniques may be reasonably proposed. But when grains with a diameter in the range of a few millimetres to a few centimetres are considered, whirling bed apparatuses can be considered. In that case the designer must have recourse to a careful study of the given problem before choosing a solution. More particularly process energy consumption as well as product quality and easy phase separation after treatment must be taken into account.

Whirling bed processes, in which high fluid flow rates are needed, often require gas recycling in order to save energy. But phase separation after contacting is easy because only two phases with widely different properties are involved. Ordinarily operations proceed as indicated in Fig. 8.

For fluidised processes in which large particles are immersed in beds of fines, an extra solid–solid separator is needed. Indeed, the two existing solid phases must be parted in order to preserve end-product quality and/or to save energy by recycling fine solids. In some cases, and especially when treating pre-packed products, separation is easy by simple screening and operation normally proceeds as shown in Fig. 9. In those cases where unpacked fissured grains or sticky surface beans are to be treated, separation may be quite difficult, if not impossible, due to the capability of the fine particles to penetrate into the large grains or to adhere to them, which may keep the designer from using the phase contacting method.

In the following sections several examples are given where direct

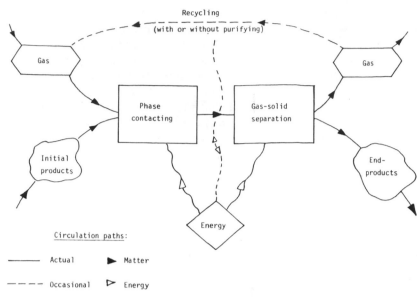

FIG. 8. Direct fluidisation processes.

and indirect gas fluidised contacting schemes have been successfully applied to various food engineering unit operations.[13,14]

3.1. Roasting

A few years ago Roltz *et al.*[15] studied coffee roasting in a traditional fluidised bed apparatus. It appears from their results that product quality is improved with regard to traditional drum processes where basically grains, enclosed in a rotating cylinder with a wall maintained at elevated temperature, are heated by direct solid–solid and wall–solid contacts. But the method is quite expensive and polluting since important quantities of hot gas are rejected.

Recently, in an attempt to save energy and to minimise pollution, coffee roasting has been tried in a whirling bed apparatus with gas recycling.[16,17] A schematic view of the experimental unit is given in Fig. 10 that shows the blower circulating hot gas through an insulated pipe loop, the electric oven used for clean gas heating, the whirling bed roasting chamber and the branch extraction circuit permitting aroma recovery from a small exit gas stream by simple cooling. To avoid too great a scatter between particle residence times, batch operating conditions were selected at first. During the loading and

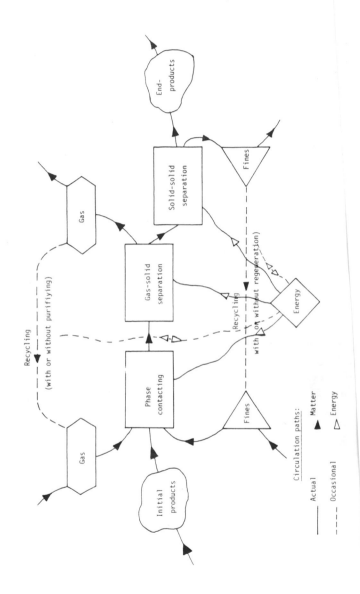

recirculating gas is low, only ordinary percentages of benzo-α-pyrene being observed on roasted grains ($\simeq 4\,\mu g/kg$) or in condensates ($\simeq 11\,\mu g/l$).

All these properties testify that the process is of value. More particularly the high gas recirculation rates that are permitted appear to be quite interesting for reducing energy consumption.

In the future, the method could be applied to other products such as peanuts, grape stones, etc.

3.2. Blanching[18,19]

Figure 11 shows a plant especially designed for blanching vegetables by direct contact with an air–water vapour mixture. Incoming products, arranged on a conveyor belt garnished with several whirling cells, are heated gradually as they go deeper into the tunnel thanks to the thermal energy removed from outgoing blanched particles by the upflow gas strea.i. In the actual 'blanching chamber' steam is introduced and temperatures as high as 90–95°C may be reached.

As indicated in Table 2, the main interests of the process lie in: low energy consumptions, because of gas recycling and heat recovery from solids; reduced waste water treatment costs, due to small condensate volumes and low COD values; and product quality improvement, as a consequence of short operating times and uniform treatment conditions.

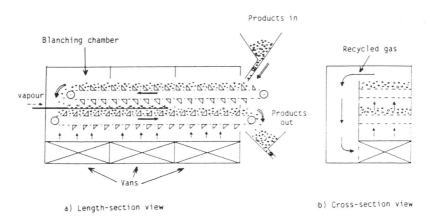

a) Length-section view

b) Cross-section view

FIG. 11. Fluidised bed blanching apparatus.

TABLE 2
COMPARISION OF TRADITIONAL AND FLUIDISED BLANCHING PROCESSES

Processes are compared on the basis of green pea treatment (1000 kg per hour)	Traditional water blanching process	Air–water vapour whirling bed process
Energy consumption	140 kW	110 kW
Polluting effluents		
volume	0·4–1·8 litre/kg	0·19 litre/kg
COD	5·9 g O_2/kg	1·2 g O_2/kg
Product quality	Washing of soluble substances	Trivial losses of vitamin C Light oxidation phenomena

3.3. Freezing

Cross-flow fluidised bed freezing of large particles transported by a conveyor belt through a pre-cooled gas stream is a well known process for Individually Quick Frozen (IQF) food production. It allows limited weight loss and in the past a wide range of fruits, vegetables and fishes have been successfully treated this way.[20–22]

More recently attempts have been made where a gas fluidised bed of fine particles kept at low temperature freezes immersed food pieces by direct contact. This solution has the advantage over the previous one of reducing gas flow rates and thermal degradation of mechanical energy, which enables the minimising of cooling requirements. On sanitary grounds the idea of using small ice particles as fines in those cases where unpacked products are to be treated has been proposed.[23,24]

3.4. Drying

Fluidisation methods have become widely used in drying technology. Fluidised bed drying has been applied not only to granular materials but also to pastes, suspensions, solutions and molten products. As a result many batch processes have been replaced by continuous processes which are much faster and more economical.[25,26]

Various kinds of fluidised beds are used in industry for solid drying: traditional beds for powders, spouted beds for large particles, vibro-fluidised beds to avoid particle agglomeration or even combined systems involving mono- or multicell arrangements. Liquids generally are dried on inert particles and the dry product is carried away from the drying chamber as a fine powder.

Recently potatoes, carrots, rice have been dried in a whirling bed apparatus.[27] In the course of experiments it has been found that puffing could be quickly operated at atmospheric pressure thanks to high gas–solid heat transfer rates prevailing in the reactor. It is worth recalling that puffing technology, ordinarily carried out in a less attractive way at reduced pressure, is useful for the manufacturing of appetizers.

In another work, drying of milk on hot large metallic beads in a multistage whirling bed column has been carried out.[28] Basically, as shown in Fig. 12, liquid milk is injected at the bottom of the column

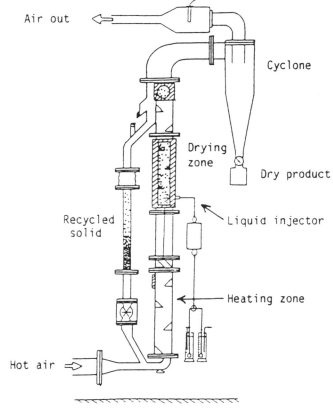

FIG. 12. Fluidised bed drying of liquids.

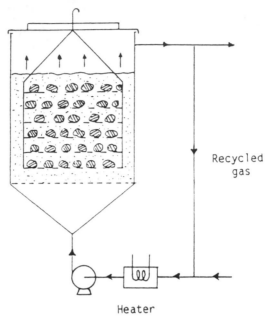

Recycled
gas

Heater

FIG. 13. Indirect fluidised bed drying of solid foodstuffs.

along with hot gas and recycled particles; then as liquid and solid go upwards drying takes place; at the top of the unit milk powder is entrained by gas to a cyclone and solids are recycled.

It is the goal of other outstanding experiments to try solid drying by object immersion in a hot bed of salt or sugar powder[29] (Fig. 13).

3.5. Atmospheric Freeze-Drying

Moisture removal from food products can be accomplished by freeze-drying which involves the sublimation of water from the frozen state directly to the vapour state. The obvious advantage of the freeze-drying process is that dehydration can be realised without exposing the product to excessively high temperatures. In addition it is normally possible to maintain the product structure in a more acceptable state, resulting in a higher quality product. The basic components of a traditional plant include an evaporator where heat is generated to be used as a source of energy for drying and a condenser which collects the vapours produced by the product. Both the condenser and the evaporator are located in a vacuum chamber.

A new and very attractive method operating at atmospheric pressure has been recently proposed[30,31] that consists of immersing the products to be treated in a gas fluidised bed of fine adsorbent particles cooled at the wall. The first step of freezing foods is realised at a very low temperature ($<$ 25°C); then solids are dried at a higher bed temperature (experiments have been carried out at up to 7°C!). In the particular working conditions of the process ice sublimation can normally take place because of the low water vapour pressure and temperature at the product surface (Fig. 14). These are due to the high rates of heat and mass transfer between the bed and the products that result from the continuous movement of adsorbents. The only energy requirements for drying are those necessary to compensate for losses

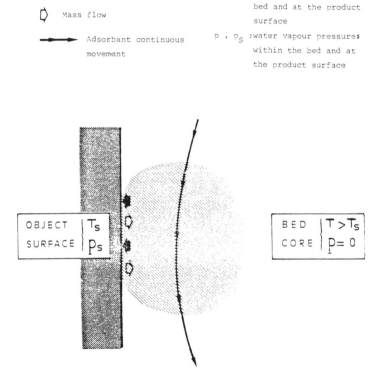

FIG. 14. Transfer mechanisms at the product surface during atmospheric freeze-drying.

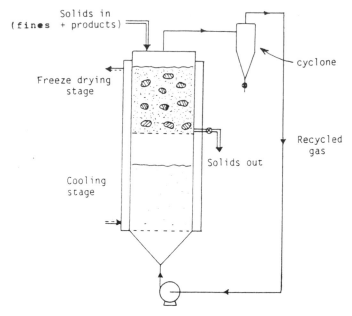

FIG. 15. Batch atmospheric freeze-drier.

and to regenerate adsorbent material from time to time. So operating costs are low. Another major advantage of this method over the traditional one is that it allows for continuous operations. A schematic view of a fluidised bed atmospheric freeze-drying unit is given in Fig. 15.

4. LIQUID–SOLID SYSTEMS

Liquid–solid applications of fluidisation in the food processing area are much more recent than the gas–solid ones. They may be classified into two types according to the physical state of foodstuffs: operations where solid grains are to be treated and operations where liquid properties must be modified to meet specific quality requirements. In the following two typical operations are presented, the first in which fluidisation has been successfully applied to solid fermentation and the second concerning a new method for liquid concentration. Then the future application of fluidised liquid–solid systems to new fields is discussed.

4.1. Fermentation of Cocoa Beans in Three Phase Fluidised Bed

Fermentation of cocoa beans is an important step in chocolate making. Traditionally the operation is worked out by heaping fresh beans on the ground and leaving them in the sun for 6 or 7 days; then the natural development of microbes controls the operation. However such a technique has typical disadvantages: the end product properties are not uniform which is detrimental to chocolate quality; large areas of ground are required which limits the use of the 'artisanal' method for treating large amounts of beans; product losses may be important when harvest time is rainy. To overcome these difficulties a new method of fermenting cocoa beans in an aqueous medium has been proposed that allows a narrow control of the main operating parameters: temperature, pH, oxygenation, microbial populations and mixing of products.[32]

Bringing cocoa beans into contact with water deeply modifies the technology of fermentation.[33] Specifically the system must be seeded twice: firstly to start the alcoholic step—in our studies a yeast strain of *Saccharomyces chevalieri* with a strong pectinolytic activity was selected—and during operation to artificially induce the following acetic step—with a bacterial strain of *Acetobacter rancens* as an example. Moreover temperature, pH, as well as aeration, must be carefully controlled and good mixing conditions must be maintained in spite of product sticky surface and high solid loads (as a general rule, liquid-to-solid mass ratio is less than one so as to minimise soluble substance extraction phenomena strongly detrimental to end-product quality).

After various unfruitful attempts carried out using either a mechanically stirred tank fermentor (strong particle damaging took place) or a traditional three-phase fluidised bed technique (there was a quick particle caking), the new three-phase fluidised bed technique described in Section 2.2.2 was proposed. Basically the reactor is a batch tank, initially filled with cocoa beans and water, and then aerated. Thanks to the special air distribution, good mixing conditions for the solids are guaranteed during the operation despite a strong juice thickening.

A very uniform and smooth treatment follows which supplies clean surface products at the end of the operation and facilitates the subsequent drying step. A schematic view of the pilot plant is given in Fig. 16, with a Roots suppressor to circulate air, a humidification column to avoid too much evaporation in the reactor, as well as regulation systems and probes for monitoring the apparatus.

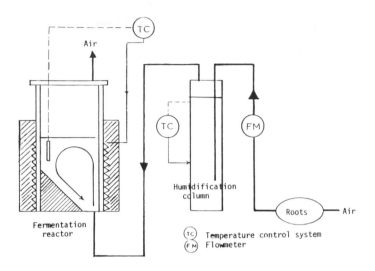

FIG. 16. Three phase cocoa bean fermentation apparatus.

In the course of the work, various operating conditions have been tested and cocoa bean processing has been judged from several criteria: microbiology (development of microbial populations), chemistry (evolution of ammoniacal nitrogen, anthocyanin pigments and volatile acids) and organoleptic properties (tasting tests on chocolate were realised using a triangular test method).

In the present state of experiments, the following conclusions may be drawn:

—through the new method, fermentation is shorter and lasts approximately 2 days;

—energy consumption for good mixing conditions in the reactor is very low (≪1 watt per kg of beans), a make-and-break aeration being enough;

—the design of overall continuous units, open to capacity improvement, is easy by connecting several reactors in parallel, with one single compressor, and staggering loading and unloading times;

—bringing the beans into contact with water does not affect organoleptic qualities adversely, in spite of chemical differences between the end-products here obtained and beans fermented in the traditional way.

So one can assert that the new technique is promising even if more experiments and working condition adjustments are needed.

4.2. Ultrafiltration of Milk in Liquid–Solid Fluidised Bed

In order to separate and to thicken colloidal solutions such as those of protein, ultrafiltration is being used more and more often in the food industry. It can reduce costs, increase product recovery, raise product purity and simplify or enhance the performances of downstream operations. However, there is a major bottleneck in this method: the so-called phenomenon of 'concentration polarisation'.

Concentration polarisation is due to solute concentrations that are higher at the membrane surface than in the bulk of the feed solution—or 'retentate'. This is caued by the fact that the rejected solutes build up concentration gradients for back diffusion into the bulk of the solution, which counteract the convective transport of solutes to the membrane. Adverse effects of the increased membrane wall concentrations are reduced quality and quantity of the separated stream—or 'permeate'. In those peculiar processes where large molecules are present (and nearly all rejected) the back diffusion velocity of them is very low; so their concentration at the interface increases and a gel layer may be formed on the membrane surface. As in scaling, this gel layer lowers the flux through the membrane considerably.

Ordinarily these undesirable effects are diminished by applying high superficial velocities to permeate in order to increase the local shear stress at the wall. Another way consists in using turbulence promoters located in the vicinity of the membrane surface. Recently the idea of choosing fluidised particles as turbulence promoters has been proposed.[34] This is very attractive at first sight since the solids might also be able to mechanically remove the gel layer when formed. However, the use of this technique necessitates the selection of membranes that are strong enough to endure the continuous bombardment of the fluid bed particles. Until recent years no material met such requirements, traditional asymmetric UF-membranes being manufactured by casting a fragile thin polymer skin (polyacrilonitrile) onto a suitable supporting layer. But in the last 2 or 3 years new materials coated with a strong mineral film (Al_2O_3, $ZrCl_4$) have come onto the market that can afford developments in this new way.

Recently work has been carried out that states the true interest of the technique.[35] A schematic view of the experimental unit is given in Fig. 17. A membrane in the form of an open tube of 27–33 mm diameter and 650 mm length, made up from alumina and manufac-

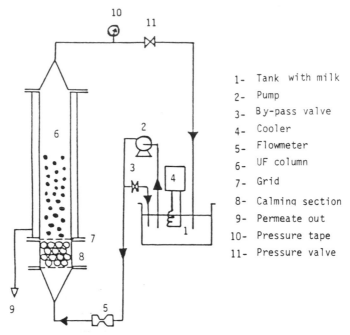

1- Tank with milk
2- Pump
3- By-pass valve
4- Cooler
5- Flowmeter
6- UF column
7- Grid
8- Calming section
9- Permeate out
10- Pressure tape
11- Pressure valve

FIG. 17. Liquid–solid fluidised bed apparatus for ultrafiltration.

tured by the French company Euroceral, was used as a filtering medium. Experiments were conducted at a temperature and a pressure of around 25°C and 1·5 bar, respectively, using reconstituted milk with 10% dry matter as fluid and 3 mm stainless steel spheres as fluidised particles. A traditional grid plate had been chosen to ensure fluid distribution at the bottom of the porous wall column. In the course of this work it has been established (Fig. 18) that: ultrafiltration flux, D, in a fluidised bed may be considerably higher than in an empty tube; membrane rejection for nitrogen compounds, T, defined as the ratio of the concentration difference between retentate and permeate to the retentate concentration, slightly decreases when passing from empty tube to fluidised bed system; but whatever the case protein yield is excellent (it is worth recalling that a high retention rate of protein is the main point of milk concentration); most favourable conditions are reached for a bed porosity around 0·68; application of fluidised turbulence promoters is advantageous with regard to energy consumption; and there is no membrane damage.

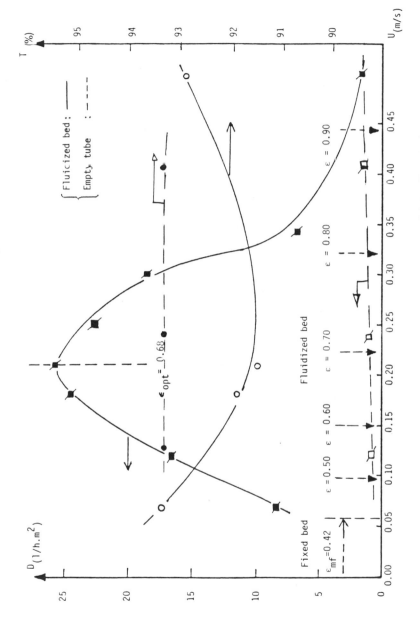

FIG. 18. Ultrafiltration flux and retention rate *vs* superficial milk velocity.

From theoretical considerations on mass transfer in homogeneous fluidised beds, the results have been interpreted. More particularly the location of a maximum flux around 0·68 has been justified. From this it may be inferred that similar properties should also be observed with fluids other than milk (blood, fruit juices, etc.).

From a practical viewpoint however it is worth noting that the use of fluidisation necessitates the selection of low circulation rates of permeate, less than the transport velocity of the chosen particles. So it may be thought that the industrial developments of the technique should be basically concerned with those processes where only low quantities of fluid are to be treated.

4.3. Other Potential Applications

The two previous examples have shown how fluidisation can help to solve new phase contacting problems such as those encountered when installing new treatment methods, or even how it can be used to improve efficiency of well-known operations. There is no doubt whatsoever that in the near future such procedures should be generalised and provide attractive solutions for other typical food engineering unit operations involving at least a continuous liquid phase, as in the following examples.

4.3.1. Extractions

The extraction of a soluble constituent from a solid by means of a solvent is generally referred to as leaching. Soluble coffee production, the manufacture of sugar, recent methods for fruit and vegetable juice making all involve processes which employ leaching on a large scale. In such processes the agitation of the solvent is important because it increases the eddy diffusion and therefore improves the transfer of material from the surface of the particles to the bulk of the solution. Moreover the better the contacting between the solid and the liquid, the greater is the rate of transfer of solubles. Sometimes solid particles are fragile and there is a strong mashing tendency, detrimental to leaching efficiency, which prohibits the use of mechanical stirrers.

A survey of the literature clearly indicates that there is no satisfactory apparatus on the market at the present time. In most industrial extractors the solvent percolates through fixed beds of coarse materials whereas the fine solids offer too high a resistance, and performances are moderate.

The characteristic properties of fluidised systems should allow the elimination of most of the impediments previously mentioned. Original apparatuses could be imagined either based on a traditional two-phase fluidised contacting method in those cases where the liquid chosen as solvent has a sufficiently low viscosity and circulates at a high enough velocity to ensure good mixing conditions, or using a well adapted gas–liquid–solid technique enabling the improvement of turbulence thanks to gas injection. In order to reach high extraction rates concentration gradients should be preferably kept at a high level by selecting multicell arrangements.

Liquid–liquid extraction that involves bringing a solution into intimate contact with a solvent in order to recover a solute (aroma, colouring substance, oil) could also be realised in an attractive way thanks to fluidisation. In that case solid particles would be used to finely disperse one liquid phase into the other in the form of droplets.

4.3.2. Freezing Concentration

The crystallisation of ice from solutions has been considered a possible means for concentrating solutions. In this operation an aqueous solution (fruit juice, wine, milk, etc.) is lowered in temperature. At a given point the freezing temperature is reached and then the solution separates into two phases, a liquid and a solid. Provided that the adherence and the inclusion of solute within the ice crystals remain small, the liquid phase has a higher concentration of solute than the original solution.

As a matter of course such an operation could be attractively worked out in a fluidised bed either in a two-phase process by cooling the liquid through the surrounding wall (but in that case a strong decrease in cooling rate might be caused by the formation of ice on the cooling surface) or in a three phase process by direct injection of a cryogenic compound into the liquid.

4.3.3. Reverse Osmosis

Obviously the results that have been presented on ultrafiltration should be applicable to reverse osmosis. However at the present time there is no membrane on the market able to endure the erosive action of moving particles.

4.3.4. Microbial or Enzymatic Reactions

The principles of solid fermentation as presented in Section 4.1 should normally be applicable to other products and more especially to the manufacture of other indigenous foods.

There are also a number of industrial food processes employing enzymes to carry out conversion—for example the hydrolysis of protein or carbohydrates, the conversion of glucose to fructose, the production of amino acids—that could be realised in fluidised beds in the future. At the present time many of these processes are worked out in a liquid medium by using either free purified enzymes extracted from microbial cells, or even whole cells that are simply 'bags of enzymes' for more complex reactions—for example the synthesis of material requiring more than one enzyme as well as an energy input. With regard to product quality and catalyst saving, a very attractive way for carrying out operations might consist in immobilising enzymes or cells on fluidised solid supports.

5. CONCLUSIONS

A lot of Unit Operations of Food Engineering can be successfully carried out using fluidised bed techniques. Contrary to widely accepted opinion, there exist several methods for realising fluidised phase contacting, and fluidisation engineering is mainly concerned with the selection of the best one according to operating cost and food quality. On that account it is worth noting that the 'whirling bed' and the 'fluidised flotation cell' techniques prove to be most favourable.

The fluid-like behaviour of solids that facilitates continuous operations, and that enables the rapid transfer of heat and mass between fluid and particles as well as high heat transfer rates to boundary surfaces, is the most important property of gas–solid systems. Because of it, fluidised beds often are used as heat exchangers, particularly where close control of temperature is required and where large amounts of heat must be added to, or removed from, the system. As an example for blanching of vegetables in an air–water vapour mixture, roasting of coffee grains, drying of biological liquids and solids, freezing of various fruits, fishes or atmospheric freeze-drying of foodstuffs.

Concerning liquid–solid applications, the food processing industry should also permit a wide development of two and multiphase fluidised bed techniques owing to the large number of Unit Operations involving at least one continuous liquid medium: liquid–liquid or liquid–solid extractions, membrane separations, freeze concentration, solid fermentation, enzymatic and microbial conversions. As a general rule for such operations fluidisation warrants an easy and smooth handling of products, an improved quality due to uniform and well-controlled working conditions, short operating times when compared to traditional methods because of increased mass transfer rates, and low operating costs.

So one can assert that all these techniques, well adapted to treat uniformly larger and larger quantities of product as well as to save more and more energy, should be of a growing importance in the next years, with a lot of industrial developments following the outstanding academic works and future research studies.

REFERENCES

1. DAVIDSON, J. F. and HARRISON, D., In *Fluidization*, Academic Press, New York, 1971.
2. BOTTERILL, J. S. M., In *Fluid-bed heat transfer*, Academic Press, New York, 1975.
3. KUNII, D. and LEVENSPIEL, O., In *Fluidization engineering*, Krieger Publishing Company, New York, 1977.
4. MATHUR, K. B. and EPSTEIN, N., In *Spouted beds*, Academic Press, New York, 1974.
5. SHAH, Y. T., In *Gas-Liquid-Solid Reactor Design*, McGraw-Hill, New York, 1979.
6. BAXERRES, J. L. and GIBERT, H., *French patent ANVAR No. 76 34 846*, 1976.
7. BAXERRES, J. L., HAEWSUNGCHARERN, A. and GIBERT, H., *Lebensm. Wissen. Technol.*, 1977, **10**, 191–96.
8. RIOS, G. M., GIBERT, H., CROUZET, J. and VINCENT, J. C., *French patent ANVAR, No. 77 15 007*, 1977.
9. RIOS, G. M., BAXERRES, J. L. and GIBERT, H., In *Fluidization*, Plenum Press, New York, 1980.
10. RIOS, G. M. and GIBERT, H., In *Fluidization*, United Engineering Trustees, New York, 1984.
11. RIOS, G. M., BAXERRES, J. L. and GIBERT, H., *French patent ANVAR No. 79 17 430*, 1979.
12. PAPACONSTANTINOU, S., ELMALEH, S. and RIOS, G. M., In *Heat and mass transfer in fixed and fluidized beds*, Hemisphere Publishing, New York, 1985.
13. HELDMAN, D. R., In *Food process engineering*, Avi Publishing, Westport, 1975.
14. CHARM, S. E., In *The fundamentals of food engineering*, Avi Publishing, Westport, 1971.
15. ROLTZ, C., MENCHUR, J. F., SANDOVAL, J. and SIMON, P., *Cafe (Lima, Peru)*, 1968, **9**, 3–18.

16. ARJONA, J. L., RIOS, G. M. and GIBERT, H., *Lebensm. Wissen. Technol.*, 1980, **13**, 285–290.
17. SHIN, H. K. and CROUZET, J., *Environ. Techn. Lett.*, 1981, **2**, 365–70.
18. BAXERRES, J. L., *Ph.D. Thesis*, USTL Montpellier, 1978.
19. RIOS, G. M., BAXERRES, J. L. and GIBERT, H., *Lebensm. Wissen. Technol.*, 1978, **11**, 176–80.
20. PERSSON, P. O., *ASHRAE J*. 1967, **6**, 42–4.
21. VAZQUEZ, A. and CAVELO, A., *J. Food Process Eng.*, 1980, **4**, 53–70.
22. NAGAOKA, J., TAKAGI, S. and HOTANIS, S. *Proc. 9th Intern. Congr. Refrig., Paris*, **2,4**, 1955.
23. MARIN, M., RIOS, G. M. and GIBERT, H., *J. Powder Bulk Solid Tech.*, 1983, **7**, 21–25.
24. LERMUZEAUX, A., *Revue générale du froid*, 1972, **11**, 1039–1102.
25. ROMANKOV, P. G., In *Fluidization*, Academic Press, New York, 1971.
26. MUJUMDAR, A. S., In *Advances in drying*, Hemisphere Publishing, Washington, 1980.
27. BAXERRES, J. L., YOW, Y. S. and GIBERT, H., *Lebensm. Wiss. Technol.*, 1983, **16**, 27–31.
28. DECLOUX, M., *Ph.D. Thesis*, INP Toulouse, 1982.
29. ABID, M., COBBINAH, S., GIBERT, H. and LAGUERIE, C., In *Drying 84*, Elsevier, London.
30. WACHET, J. N., *Ph.D. Thesis*, USTL Montpellier, 1978.
31. KAMAL, Y., *Ph.D. Thesis*, INP Toulouse, 1984.
32. JACQUET, M., VINCENT, J. C., RIOS, G. M. and GIBERT, H., *Café, Cacao, Thé (France)*, 1981, **1**, 45–54.
33. JACQUET, M. and VINCENT, J. C., *French patent ANVAR No. 79 33 054*, 1979.
34. VAN DER WAAL, M. J., VAN DER VELDEN, P. M., KONING, J., SMOLDERS, C. A. and VAN SWAAY, W. P. M., *Desalination*, 1977, **22**, 465–83.
35. MONTLAHUC, G., RIOS, G. M. and TARODO, B., *Lebensm. Wissen. Technol.*, 1984, **17**, 185–90.

INDEX

3-A Sanitary Standards
  cleaning fluid flow velocity, 106
  food plant hygienic design principles, 111–12
12D process, in canning, 5, 6

Aerobic plate count (APC), effect of sodium nitrite, 70, 71
Ageing effects, food soil, 109–10
Aluminium, as material of construction, 113
Ancillary equipment, hygienic design, 117
Apple juice
  multiple-effect falling-film evaporator plant, 220–32
    feed pre-heating, 227–31
    liquor re-circulation, 231–2
    optimal number of effects, 222–7
    physical properties, regression equations for, 222
Aroma, addition to microwave-cooked beefsteaks, 163, 165
Arrhenius equation, 9, 10
  thermal–time predicted by, 269
Arrhenius thermal inactivation model, 251, 253
Ascorbate, effects on packaged meats, 75–6
Ascorbic acid, oxidation in fruit juices, 221
Aspergillus niger, effects of microwave heating, 159, 160
Atmospheric freeze-drying processes, fluidised-bed methods, 292–4

Atypical streptobacteria, in meat products, 50, 67

Bacillus cereus
  growth inhibited by CO_2, 64
  low-temperature growth, 53, 57
  salted meat products, in, 69
  sodium nitrite, effects of, 73
Bacillus stearothermophilus, effects of microwave heating, 159
Bacillus thermosphacta
  meat products, in, 50, 51
  nitrite, effect of, 71
  phosphates, effect of, 81
Bacteria
  dielectric heating, effects of, 139, 154–61
  microwave heating, effects of, 155–7, 159–61
  minimum relative humidity for growth, 129
  see also Individual species by name:
    Bacillus spp.; *Clostridium botulinum*; *Pseudomonas* spp.; *Salmonella* spp.; *Staphylococcus aureus*; *Yersinia enterocolitica*
Bacterial materials, food soil formation affected by, 98–9
Bacteriological process evaluation, canning processes, 8
Baking processes
  combined (conventional/infra red/microwave) methods, 166, 170

Baking processes—*contd.*
dielectric heating applications, 166–8, 170
Beefsteaks, microwave-cooked, 165
Biscuits, microwave baking of, 168
Blanching processes
dielectric heating methods, 161–2
fluidised-bed methods, 289–90
Bologna-type sausage, 47, 54, 55, 56, 60–1, 77
Botulism, *see also Clostridium botulinum*. . .
Botulism outbreaks, canning process effects, 6, 13
Bread, baking by microwave, 166–8
Browning reactions
apple juice, 221
kinetic data for, 254
microwave-cooked beefsteaks, absence in, 165
Burning-on, food soil, 97, 104
Butter
fat separation using dielectric heating, 142
microwave defrosting of, 174

Carbon dioxide
meat products, in, 58–9
packaged meat bacteria growth affected by, 62–6
Casolari thermal inactivation model, 252, 253
Cheese processing, dielectrically heated plant for, 141–2
Chocolate
melting, by dielectric heating, 142
see also Cocoa beans
CIP systems, 117–22
advantages, 117
centralised, 117–18
multi-use systems, 122
re-use systems, 120–1
rotating spray turbines for tanks/vessels, 120
satellite systems, 118–19
single-use systems, 120, 121
sprayballs for tanks/vessels, 119

Cleaning, food processing plant
age-of-deposit effects, 109–10
contact time effects, 107–8
detergents, effects of, 108–9
practical applications, 111–22
problems, 97
surface finish effects, 110–1
temperature effects, 103–5
turbulence effects, 105–7
Cleaning-in-place (CIP) systems, 117–22; *see also* CIP systems
Cleanliness, evaluation techniques for, 99–103
Climbing-film evaporators, compared to falling-film evaporators, 194
Clostridium botulinum
kinetic data for thermal inactivation, 254
nitrite, effects of, 72–3
pH effect on, 18
phosphates, effect of, 82
salted meat products, in, 68
sorbates, effect of, 77
thermal destruction, 18
Clostridium perfringens, low-temperature growth, 53, 57
Cocoa beans
fermentation in fluidised bed system, 295–7
flow diagram, 296
high-frequency roasting of, 151, 152–3
economics, 153
see also Chocolate
Coffee
fluidised-bed roasting, 286, 288–9
microwave roasting of, 154
Coliform bacteria, effects of sodium nitrite, 73–4
Commercial sterility
definition, 4
factors affecting, 4
Computer models
plate heat exchanger, 267
scraped surface heat exchanger, 265–7
tube heat exchanger, 265

Computer programs
  dielectric characteristics of
    foodstuffs, 131
  microwave heating, 133–4
  thermal processing lethality cal-
    culation, 39
Contamination, sources of, in meat
  products, 46–8
Cooked-meat products
  contamination of, 46–7
  pH effects on spoilage, 80–1
Cooking processes
  dielectric heating applications,
    162–6, 169–71
  fluidised-bed methods, 274, 286–9
  see also Roasting
Corn, dielectric characteristics of, 131
Count reduction system, 8
Cross-contamination, meat products,
  48
Cube sugar, dielectric drying of, 147
Cured meat products
  contamination of, 46, 47
  gas packaging of, 66
  pH effects on spoilage, 80
  post-packaging pasteurisation of,
    48

D–Z thermal inactivation model,
  251, 252, 253
Decimal reduction time (D) values
  definitions, 5, 251
  typical values, 18
Defrosting, dielectric heating applications, 171–5
Detergents
  concentration effects on soil
    removal, 109
  definition, 108
  formulation effects on soil removal,
    108–9
DFD (dark–firm–dry) meat, 78
  microbial spoilage of, 78–9
Dielectric characteristics, foodstuffs
  measurement, 130–2, 134
  reasons for variability, 130
Dielectric heating, 129–36

Dielectric heating—contd.
  advantages, 175–6
  applications, 140–75
  automatic control of, 170
  baking applications, 166–8, 170
  blanching applications, 161–2
  chemical processes in, 137–8
  cooking applications, 162–6,
    169–71
  critical energy density, 135
  defrosting applications, 171–5
  drying applications, 144–50
  effects of, 136–40
  heating applications, 140–4
  mathematical model, 134
  micro-organisms affected by, 139,
    154–61
  microstructure, importance of, 137
  model studies for air drying, 147–8
  pasteurisation applications, 154–5
  pay-back period, 176
  penetration depth as function of
    frequency, 136
  power absorbed, 129–30
  roasting applications, 150–4
  sterilisation applications, 155–61
  trapped moisture, 136–7
Direct measurement, temperature/
  time profiles, 10, 23–9
Drying process, dielectric heating,
  144–50
Drying processes, fluidised-bed
  methods, 290–2

Eddy diffusion processes
  fluidised beds, in, 300
  heat transfer, in, 254–5
Egg mixtures
  microwave cooking of, 163
  microwave defrosting of, 172
  microwave vacuum drying of, 149
Enterobacteriaceae
  salted meat products, in, 67–8
  sodium nitrite, effects of, 72
Enzymatic reactions, fluidised-bed
  methods, 301–2
Equivalent heating times, 249
  milk thermal processing, 248

Ergun equation (for fluidisation), 276
Euroceral ultrafiltration membrane, 297–8
Extraction processes, fluidised-bed methods, 300–1

F values, 5, 6, 7
Falling-film evaporators (FFEs)
  advantages, 190, 193
  characteristics, 195–6
  compared to climbing-film evaporators, 194
  fouling of heat transfer surfaces, 194
  general description, 191–3
  heat transfer in, 197–209
    models, 204
    overall coefficient, 202–9
  nomenclature, 184–90
  performance characteristics, 194–9
  types, 192, 193
  see also Multiple-effect falling-film evaporators
Falling-films
  flow parameters in, 197–9
  laminar flow heat transfer, 202
  local heat transfer coefficients, 197–202
  turbulent flow heat transfer, 200–1
Fat-melting plant, dielectrically heated, 141
Fermentation processes, fluidised-bed methods, 295–7, 301–2
Film permeability
  CO_2/O_2 ratio, 58–9
  packaged meats, 58–62
  temperature effects, 60
Fish, microwave defrosting of, 171, 174
Flow parameters, falling-films, 197–9
Flow velocity effects, food soil removal, 106–7
Fluid flow
  falling films, in, 197–202
  food soil removal, effects on, 106–7
  heat exchangers, effects in, 261–4
Fluid models, 260
  Ellis model, 260

Fluid models—contd.
  Eyring Powell model, 260
  Ostwald–de Waele model, 260
  Power law model, 260
Fluidisation
  advantages/disadvantages of, 277–9
  applications, 274, 286, 288–302
    blanching, 289–90
    cocoa bean fermentation, 295–7
    drying, 290–2
    extraction, 300–1
    fermentation, 295–7, 301–2
    freeze-drying, 292–4
    freezing, 290
    freezing concentration, 301
    milk drying, 291–2
    milk ultrafiltration, 297–300
    reverse osmosis, 301
    roasting, 286, 288–9
    ultrafiltration, 297–300
  direct techniques, 279, 281–2, 284–5, 286
  gas–solid systems, 285–94
  indirect techniques, 282–4, 287
  liquid–solid systems, 294–302
  new systems, 281–5
  phenomena, 275–7
  practical limitations, 280–1
  velocity range, 276
Fluidised beds
  comparison of gas–solid and liquid–solid systems, 279
  factors affecting, 279
  fixed-bed, 275, 277
  flotation cell technique, 282–3
  gas–solid systems, 275, 276, 281–4, 285–94
  liquid–solid systems, 275, 276, 284–5, 294–302
  pressure drop over, 277
  spouted beds, 280, 281
  three-phase, 277, 278, 295–6
  types, 275–7
  whirling bed processes, 281–2, 284–5
Formula Methods, thermal process determination, 29–37, 252–3

Frankfurters
gas packaging, 66
heat treatment, 47, 48
Freeze-drying processes
dielectrically heated vacuum drying
compared to, 149–50
fluidised-bed methods, 292–4
transfer mechanisms in, 293
Freezing concentration processes,
fluidised-bed methods, 301
Freezing processes, fluidised-bed
methods, 290
Frozen foods, dielectric characteristics of, 133
Frozen products, microwave defrosting of, 171–5
Fruit juices
high-frequency pasteurisation of,
154–5
see also Apple juice
Frying processes, combined with
microwave heating, 170
Fungi, effects of microwave heating,
159, 160

Galileo number, 276
Gas packaging, effects on packaged
meats, 62–66
General Method, thermal process
determination, 24, 28–9, 38
Gigavac drying system, 149
Glass
containers for microwave-sterilised
foodstuffs, 171
plant construction, as material of,
113
Glucose, effect on spoilage of meat
products, 80
Greening, of meat products, 49, 78

Heat exchangers
fouling of surfaces, 97, 194
heat/flow models, 261–4
Lagrangian formulation, 263–4
laminar flow, 264
reactor models, 262–3

Heat exchangers—contd.
heat/flow models—contd.
turbulent flow, 262–3
nomenclature, 245–7
see also Plate. . . ; Scraped
surface. . . ; Tube. . .
Heat penetration tests, 24–5
statistical approach, 25
Heat transfer coefficients (HTCs)
local in falling films, 197–209
overall in falling-film evaporators,
202–9
Heat transfer models
can temperature profiles predicted
by, 10
direct measurement analysis, 23–9
can-to-can variations, 27
complex heat transfer effects,
25–6
retort-batch variations, 27–8
temperature measurement
errors, 26
Formula Methods for, 29–40
survival number determined from,
13–15
Heat treatment
meat micro-organisms affected by,
48–9
microbial counts affected by, 46–7
see also Thermal processing
Herbs and spices, microbial counts
on, 46
High-frequency electromagnetic
fields, heating by, 129–36
applications, 140–75
effects, 136–40
High-temperature short-time (HTST)
process
falling-film evaporators, 192
principles, 248–9
sterilisation plants compared with,
247
HI-VAC packaging, 59
Hops, dielectric drying of, 145, 146
Hygienic design
ancillary equipment, of, 117
basic principles, 111–12
materials of construction, 112–13

Hygienic design—*contd.*
  pipelines and fittings, of, 115–16
  storage tanks and vessels, of,
      113–15

Infra-red heating, combined with
    microwave heating, 166
Inoculated pack system, 8

Kamboko gels, dielectrically heated
    plant for, 141
Kapitza theory, film flow, 198
Kinetic models
  application of, 252–4
  combined concentration/
      temperature effects, 251–2
  concentration effects, 250–1
  spore destruction in thermal
      processing, 16–19
  spore survival in thermal process-
      ing, 13–15
    conduction-only model, 13–14
    convection-only model, 13
  temperature effects, 251
  thermal processing, inadequacy of,
      38

Lactobacilli, packaged meat prod-
    ucts, 51, 54–7
Lagrangian formulation, heat ex-
    changer fluid flow, 263–4
Lethality calculations
  kinetic models, 9, 13–23
  overall method, 10–12
  principles, 8–13
  temperature effects on, 9–10
Lethality (*F*) values, 5, 6, 7
Light, meat products affected by,
    82–3
Lipids, food soil formation affected
    by, 98
Liver, microwave defrosting of, 171,
    173
Luncheon meat
  contamination of, 47
  heat treatment of, 48

Maillard reactions: *see* Browning
    reactions
Malt, dielectric drying of, 151
Meat
  fresh, compositional effects, 78–80
  micro-organisms in, 49–52
  microwave defrosting of, 171, 175
  microwave heating of, 144, 163,
      165
  packaged, refrigerated storage of,
      45–85
  reheating by microwave, 169
Meat products
  compositional effects, 78–82
  fatty portions, microbial growth,
      83–4
  lighting effects, 82–3
  pH effects, 78–81
  phosphate effects on spoilage of,
      81–2
  see also Packaged meats
Microbial reactions, fluidised-bed
    methods, 301–2
Micrococcaceae, in salted meat
    products, 66–7
Micro-organisms
  kinetic data for thermal inactiva-
      tion, 254
  microwave heating, effects of,
      155–7, 159–61
  thermal inactivation of, 252–4
Microprocessor-based microwave
    heating control systems, 170
Microwave freeze drying systems,
    149–50
  economics, 150
Microwave ovens, domestic, micro-
    organism survival in, 160
Milk
  fluidised-bed drying, 291–2
  fluidised-bed ultrafiltration, 297–
      300
  time–temperature relationships,
      248, 249
Mineral salts, food soil formation
    affected by, 98
Moulds, minimum relative humidity
    for growth, 129

Multiple-effect falling-film evaporators (MEFFEs)
apple juice production, 220–32
backward flow calculations, 224, 225–7, 229, 230, 231–2, 233, 234
cost analysis, 218–19, 221, 237–44
design
calculation procedure, 211–17
details, 237–44
optimisation procedure, 217–20
falling-film evaporator design, 237–8
feed pre-heating, effect of, 227–31
flow diagrams, 208, 210
input data required, 220
interstage pumps, 225
design and cost analysis, 240–2
liquid–vapour separator design, 238–40
liquor recirculation, 231–2, 233
mathematical models, 209–11
modelling, 209–17
optimal number of effects in design of, 222–7
vacuum system design, 242–4

Navier–Stokes fluid flow equations, 259
Nematospora infection, microwave treatment of, 161
Nitrates, effects on packaged meats, 74–5
Nitrites
packaged meats, effects on, 70–4
sodium chloride, effect of, 70
sodium isoascorbate, effect of, 76
Nitrogen, bacterial growth inhibited by, 64, 65
Nomenclature
falling-film evaporators, 184–90
heat exchangers, 245–7
thermal process determination, 1–3
'Non-square' processes, 22
Nusselt film flow model, 198, 201
Nutritional constituents
falling-film evaporators, effects on, 221, 227

Nutritional constituents—contd.
microwave cooking, effects on, 162–3, 166
Nuts
dielectric characteristics, 131–2
microwave roasting, 151, 154

Packaged meats
ascorbate effects, 75–6
contamination, sources of, 46–8
cooked/cured, 51–2
film permeability for, 58–62
gas packaging effects, 62–66
microbial interactions in, 54–8
micro-organisms in, 50–2
temperature effect on, 52–4
microwave heating, 166
nitrate, effects of, 74–5
nitrite, effects of, 70–4
oxygen concentration, effects of, 62
refrigerated storage, 45–85
rewrapping, 65–6
sodium chloride, effects of, 66–70
sorbate/sorbic acid, effects of, 76–8
Parisian sausages, dielectric cooking of, 163, 164
Pasta products
dielectric characteristics, 132
dielectric drying, 145–7, 149
high-frequency sterilisation, 156
Pasteurisation
dielectric heating processes, 154–5
meat micro-organisms affected by, 47, 48–9
post-packaging, 48–9
Pastry products, high-frequency sterilisation of, 157, 159
Paté, dielectric cooking of, 163
Pathogenic bacteria
microwave heating, effects of 155–6
packaged meats
lactic acid bacteria effects, 56–7
nitrite effects, 72–3
temperature effects, 52–4
vacuum packaging film effects, 61

Peanuts, dielectric roasting of, 151, 154
pH effects
    cooked-meat product spoilage, 80–1
    cured-meat product spoilage, 80
    fresh-meat spoilage, 78–80
    thermal resistance of spores, 18
Phosphates, effect on spoilage of meat products, 81–2
Pipelines
    cleaning, 120–2
    hygienic design, 115–16
    International Dairy Federation (IDF) joints, 115, 116
    recessed ring type (RJT) joints, 115, 116
    volumetric capacity, 118
Plastic films
    permeability, 58–62
    plasticiser migration during micro-wave heating, 166
Plate heat exchangers
    cleaning, 97, 104–5, 107
    falling-film evaporator feed pre-heating, 228
    residence time distributions, 267, 268
Poultry
    feather release by microwave technique, 169
    microwave heating, effects of, 161, 165, 169
Pre-cooking processes: see Blanching processes
Process evaluation, and sterility confirmation, 7–8
Processing plant
    cleaning, 95–122
    hygienic design principles, 111–17
Protein denaturation
    food soils, in, 96, 98
    kinetic data for thermal effects, 254
    microwave heating, 162
Pseudomonas spp.
    effect of phosphates on growth, 82
    growth inhibited by CO_2, 63, 64
    pH effects on growth, 78, 79

Q values, definition of, 251

Radio frequency heating systems, baking process, 168
Random-walk procedures, heat exchanger fluid flow, 263–4
Rapeseed oil extraction, using dielectric heating, 143–4
Raytherm microwave defrosting system, 175
Ready-made meals, microwave defrosting of, 174
Recirculation falling-film evaporators (RFFEs)
    apple juice production, 231–2, 233
    characteristics, 195–6
    schematic diagram, 193
Relative humidity, minimum for micro-organism growth, 129
Reverse osmosis, fluidised-bed methods, 301
Reynolds' equations, 258
Reynolds number, 259, 276
Roasting processes
    dielectric heating applications, 150–4
    fluidised-bed applications, 286, 288–9

Salami, microwave defrosting of, 174
Salmonella spp.
    low-temperature growth, 53, 57
    microwave heating, effects of, 155–6
    potassium sorbate, effect of, 77
    salted meat products, in, 69
    sodium nitrite, effect of, 73
    vacuum-packaging film, effects of, 61
Salt, effects on packaged meats, 66–70
Sausages
    dielectric cooking, 163, 164
    nitrites in, 71, 72, 73
    sorbates in, 77
    see also Bologna-type sausages; Frankfurters

Scale-up, canning retorts, 28
Scrambled eggs, dielectric cooking of, 163
Scraped surface heat exchangers
  residence time distributions, 265–7, 268
  thermal–time distributions, 267, 269–70
Seeds, dielectric characteristics of, 132
Single-pass falling-film evaporators (SPFFEs)
  characteristics, 195–6
  feed pre-heating, effects of, 228–30
  schematic diagram, 193
Skinner Company, commercial falling-film evaporators, 191
Sliced meat products
  contamination, 47
  gas packaging, 63, 64
  sorbate effects on, 77
Sodium ascorbate, effects on packaged meats, 75–6
Sodium chloride, effects on packaged meats, 66–70
Sodium nitrite, effects on packaged meats, 70–4
Soil (food deposit)
  causes, 95
  characteristics, 96
  formation mechanisms, 97–9
  removal
    factors affecting, 103–111
    reasons for, 96
    temperature effects on, 103–5
  temperature effects
    deposition, on, 97
    removal, on, 103–5
Sorbates, effects on packaged meats, 76–8
Soya beans, dielectric heating of, 142–3, 148–9
Spoilage, and canning sterility, 6–7
Spore destruction, kinetic models for, 16–19
Spore survival, kinetic models for, 13–15
Spores, thermal resistance of, 17–19

'Square' processes, 21
Stainless steel, as material of construction, 112–13
  surface finish of, 110–11, 116
Staphylococcus aureus
  low-temperature growth, 52–3, 56–7
  microwave heating, effects of, 156, 157
  potassium sorbate, effect of, 77
  salted meat products, in, 69
  sodium nitrite, effects of, 73
Sterilisation, dielectric heating processes, 139, 155–61
Sterility
  definitions, 4–5
  safety levels, 6
Stokes–Einstein diffusivity equation, 261–2
Storage tanks/vessels
  cleaning, 119–20
  hygienic design, 113–15
Sugar cubes, dielectric drying of, 147
Sunflower seeds, dielectric heating of, 143
Surface cleanliness evaluation techniques, 99–103
  bacteriological methods, 101–2
  electrical conductivity methods, 103
  fluorescent dye method, 100–1
  light-scattering technique, 101
  light-transmittance techniques, 101
  optical methods, 100–1
  radioactive tracer techniques, 102–3
  squeegee–floodlight test, 100
  turbidimetric methods, 101
  visual methods, 99–100
  weighing methods, 100
Surface finish effects, food soil removal, 110–11

Temperature effects
  canning lethality calculations, 9–10
  film permeability, 60
  food soil deposition, 97

Temperature effects—*contd.*
  food soil removal rates, 103–5
  kinetic data, 254
Temperature models, thermal pro-
    cessing, 19–23
  'non-square' processes, application
    to, 22
  'square' processes, application to,
    21–2
Temperature/time profiles
  direct measurement, 10, 23–9
  Formula Methods, 10, 29–40
  milk thermal processing, 248, 249
Thermal conductivity, food materials,
    129
Thermal process determination, 1–40
  Formula Methods, 29–37, 252–4
    application range, 29–30
    Ball and Olson's method, 36–7
    Ball's method, 33–4
    batch-to-batch variations, 30–1
    can-to-can variations, 30–1
    computer programs, 37, 39
    data, 39
    derivation problems, 31–2
    finite difference methods, 37
    Gillespy's methods, 34, 36
    Hayakawa's method, 34–5
    mass average survivor deter-
      mination methods, 35–7,
      38
    single critical point methods,
      32–5, 38
    Steele and Board's method, 35
    Stumbo's methods, 34, 36
  General Method, 24, 28–9, 38
  nomenclature, 1–3
Thermal processing
  general description, 3–4
  milk, 248, 249
  product changes during, 248, 249
    kinetics of, 249–54
Thermal–time distributions (TTDs),
    262
  comparative results for various
    heat exchangers, 269–70
  concept, 262

Thermal—time distributions
    (TTDS)—*contd.*
  Lagrangian formulation to estab-
    lish, 263–4
  laminar flow, 24
  reactor models for determination
    of, 262–3
  turbulent flow, 262–3
  usefulness of concept, 271
Thermally accelerated short-time
    evaporation (TASTE) eva-
    porators, 192
  disadvantages, 192
Time–temperature relationships, 10,
    23–40
  milk thermal processing, 248, 249
Tobacco products, dielectric drying
    of, 145
Trichinella spiralis, effects of
    microwave heating, 159
Tube heat exchangers
  residence time distributions, 265,
    268
  thermal–time distributions, 265,
    269–70
Two-pass falling-film evaporators
    (TPFFEs)
  schematic diagram, 193

Ultrafiltration processes, fluidised-
    bed methods, 297–300
Ultra-high-temperature (UHT) pro-
    cess, principles of, 248–9
Underprocessing
  meat micro-organisms affected by,
    49
  reasons for, 12–13

Vacuum drying systems, dielectrically
    heated, 149–50
Vacuum Foods Company, first
    commercial falling-film eva-
    porator, 191
Vacuum-packed meat products
  film permeability effects, 59, 60
  microbial interactions in, 55–6
  spoilage bacteria in, 51, 52

Veal, microwave vacuum drying of, 149
Vegetables
  fluidised-bed drying, 291
  microwave drying, 148
  microwave heating, 160, 162
Vitamin destruction
  apple juice, 221
  microwave heating, 162
  temperature-induced, kinetic data for, 254

Water
  constituent of food, as, 128
  minimum for growth of micro-organisms, 128–9
  relative dielectric constant, 130

Whirling (fluidised) beds, 281–2, 284–5

Yeasts
  biomass drying by dielectric heating, 148
  minimum relative humidity for growth, 129
  salt tolerance, 68
Yersinia enterocolitica
  growth inhibited by lactobacilli, 57
  low-temperature growth, 53–4
  salted meat products, in, 70
  sodium nitrite, effects of, 74
  vacuum-packaging film effects, 61

Z values, definition of, 251